Cities and Nature

Series Editors

Peter Newman, Sustainability Policy Institute, Curtin University, Perth, WA, Australia

Cheryl Desha, School of Engineering and Built Environment, Griffith University, Nathan, QLD, Australia

Alessandro Sanches-Pereira, Instituto 17, São Paulo, São Paulo, Brazil

Cities and Nature fosters high-quality multi-disciplinary research addressing the interface between cities and the natural environment. It provides a valuable source of relevant knowledge for researchers, planners and policy-makers. The series welcomes empirically based, cutting-edge and theoretical research in urban geography, urban planning, environmental planning, urban ecology, regional science and economics. It publishes peer-reviewed edited and authored volumes on topics dealing with the urban and the environment nexus, including: spatial dynamics of urban built areas, urban and peri-urban agriculture, urban greening and green infrastructure, environmental planning, urban forests, urban ecology, regional dynamics and landscape fragmentation.

More information about this series at https://link.springer.com/bookseries/10068

Nicholas Ardill

Growing Food in Cities

Social Innovation Strategies for Sustainable Development

Springer

Nicholas Ardill
Portsmouth School of Architecture
University of Portsmouth
Portsmouth, UK

ISSN 2520-8306 ISSN 2520-8314 (electronic)
Cities and Nature
ISBN 978-3-030-98474-8 ISBN 978-3-030-98475-5 (eBook)
https://doi.org/10.1007/978-3-030-98475-5

This Springer imprint is published by the registered company Springer Nature Switzerland AG
The registered company address is: Gewerbestrasse 11, 6330 Cham, Switzerland

To my mother Anne, an inspiring educator and environmentalist

Preface

We are living in a time where there is an imperative to expand social action and environmental strategies for tackling local to global challenges. The process of transformation towards realising sustainable development will demand innovative solutions. If nature-based solutions can be diffused and increased in coverage and connectivity, they could make significant steps in helping our communities to come together and create more inclusive, sustainable cities.

Recognising the opportunity for change, I explore in this book the process of 'social innovation' seen as innovation for the public good. It concerns human agency, environmental dynamics, and socially valuable spaces. My research will reveal how innovation progresses in urban food growing spaces aligned with wider social goals. This study sits within the convergence of current planning thought on the significance of urban food systems, social innovation, and sustainable development.

Urban food growing in various forms has multiplied in recent times. This green and inclusive urbanism creates social value for the health, wellbeing, and welfare of local inhabitants. Therefore, there is a convincing argument to investigate novel spatial practices that can enable cities to meet the needs of an increasing population. Despite the mounting interest in collective approaches to sustainable development, limited attention has been given to the diverse ways in which this social action has been pursued. How are urban food growing spaces produced through social innovation? What are the innovative processes that can be translated in a replicable model to other cities, yet suitable for local needs, to support the development of healthier, more socially just built environments?

Drawing on the latest research I provide timely insight into the agency of social innovation in urban food spaces. While it is recognised that co-operative processes may support the development of planning principles in the making of sustainable cities, cross-sectoral approaches to innovation are routinely absent in contemporary discussions. This book aims to advance understanding of social innovation by examining its processes from the perspective of different levels of urban stakeholders. It probes the gap between what we know and how bottom-up, top-down, and hybrid approaches can function in the production of the built environment to develop a process model of social innovation in urban food growing.

This book is an outcome of a recently completed research project: 'Emerging Places of Social Innovation (POSI). The co-production of space between multilevel stakeholders: the case of productive urban green infrastructure', as undertaken by the author from 2015 to 2020. It comprises six parts with the case studies on social innovation in urban food spaces forming the main body of this publication.

Part I, *Framing the Research*, introduces the concept of social innovation, and outlines what it is, who is doing it, and why it is significant to satisfying human needs and tackling societal challenges, such as health and wellbeing, social justice, food security, and climate change. Chapter 1 describes how little attention has been paid to the process of social innovation in the production of urban spaces and proposes how the topic might be investigated, whilst explaining what this will offer for the study of sustainable development. Chapter 2 sets out the research design and case study strategy, including details on the principal research sites, the study participants across three organisational levels, and the methods used to collect data.

Part II, *Social Innovation in Urban Spaces: A Conceptual Framework*, reviews present understandings in order to integrate the findings on the spatial characteristics of social innovation, co-operative processes, and environmental dynamics. Chapter 3 examines the existing research on social innovation for sustainable development, identifying the area of the collaborative production of space between bottom-up to top-down urban stakeholders as a way to examine this social action. It highlights a lack of understanding in how different contexts and stakeholders' influence collaborative processes and place-based outcomes. Chapter 4 draws on insights on social innovation and process dynamics to construct a conceptual framework to be tested and developed in this study.

Part III, *Case Studies of Social Innovation in Urban Food Spaces*, explore social innovation processes in the production of common spaces for food growing in cities, both within the UK and across Europe. Chapter 5 presents an overview of European activities around social innovation in collaborative spatial development by drawing on case studies from 12 European cities from all four European geographic regions to enable a comparative view of practice. Chapters 6 to 11 provide an in-depth analysis of six case studies across the cities of Brighton & Hove and Portsmouth, UK from the perspective of different organisational levels of urban stakeholders.

In the first of the three Brighton & Hove located cases, Chap. 6, 'The Bevy Edible Garden', is a resident led development that transformed the site of a former public house into a community owned hub and social enterprise; comprising several cultivation spaces interlinked with a community cafe and training kitchen. A holistic approach was adopted to support community development while multiple events and activities engaged residents, including the formation of a community co-operative with over 800 members. Chapter 7 examines 'Racehill Community Orchard (Harvest)', where a Brighton & Hove-based food advocacy organisation partnered with permaculture specialists to develop the largest community orchard within the city at 1.30 ha. As an exemplar project, it served as a template for 'Harvest Brighton & Hove', an urban development programme that produced over 50 food growing spaces citywide. The third case study analysed, Chap. 8, is 'The Keep Community Orchard (PAN06)'. This Brighton & Hove City Council development was integrated with a public archive centre for East Sussex. As a precedent for urban planning, the aim was to test innovative approaches in helping the city find ways to transition towards sustainable development. The project influenced the adoption of planning policy to support urban food growing.

In the first of the three Portsmouth located cases, Chap. 9, 'Southsea Greenhouse' is a grassroots development that started at a seafront concession before transforming wasteland within a park into a cultivation space and community run social enterprise. Local residents formed a co-operative with over 200 members and sold shares in the community owned venture. Chapter 10 examines 'Charles Dickens Orchard Trail', where a Portsmouth-based environmental organisation developed a series of fruit orchards along a heritage walking trail to connect deprived neighbourhoods within a highly urbanised centre. A co-design process with the local authority assisted its planning and design. The final case study, 'Stacey Community Orchard (Healthy Pompey)', is examined in Chap. 11. This Portsmouth City Council development transformed a neglected area surrounding a community centre into a demonstration space for food growing in small urban spaces. Central Government funding was obtained from a UK nationwide 'Healthy Towns' experimental programme to trial innovative ways of changing resident behaviour to live healthier lives, leading to the development of 11 community food growing spaces across Portsmouth.

Part IV, *Social Innovation Processes, Patterns and Contextual Forces in Urban Food Spaces: A Cross-Case Analysis*, draws together and combines findings from across the case studies. Chapter 12 analyses the relationship between key innovation processes to establish a common pattern of social innovation in spatial production. Chapter 13 identifies the contextual influences on innovation agency across three spatial scales: neighbourhood, urban system, and national system.

Part V, *Social Innovation Model in the Production of Urban Food Spaces*, develops the study's conceptual framework into a dynamic explanation of spatial production. Chapter 14 advances current understanding of social innovation by formulating a process model for the production of urban food spaces. The sequence of innovation processes within this model is shown to have application to cities both within the UK and across Europe.

Part VI, *The Progression of Social Innovation in Urban Food Spaces*, discusses emerging key insights from this book and conclusions are drawn on their broader meanings and implications. Chapter 15 reflects on the research aims and questions and advances the current understanding of social innovation in spatial production for sustainable development. Chapter 16 concludes the book with recommendations for policymakers and practitioners to support social innovation in urban food spaces.

In summary, this book examines social innovation strategies in the collaborative development of spaces for growing food in cities. It enables readers to gain valuable insights into an innovative social and spatial practice whilst advancing knowledge in an emerging area of research. The book will also be of great relevance to social activists, urban designers, planners, and decision-makers with an interest in applying this expertise to their own neighbourhoods and cities.

Portsmouth, UK Nicholas Ardill

Acknowledgements

This book would not have been possible without the support from some outstanding academics. I am indebted to Fabiano Lemes de Oliveira, for his knowledge, inspiration, and assistance in the field of socio-ecological research. Also, to Pablo Martínez Capdevila who provided encouragement and guidance, and to Silvio Caputo, Steffen Lehmann, and Oren Lieberman for their constructive insight. A thank you also goes to the peers that I have met at the conferences I have attended, which have contributed to interesting discussions in my field.

I would like to express my gratitude to the varied and diverse social innovation practitioners who agreed to participate in this research. I am thankful for your time and consideration during the process and your enthusiasm to discuss and learn with me. In addition, I am indebted to the wider network of peers within Brighton & Hove and Portsmouth, and in the European cities I have visited who added to this research.

My special thanks to Juliana Pitanguy, Publishing Editor at Springer for her enthusiasm to bring this book to a wider audience and to Peter Newman, Cheryl Desha, and Alessandro Sanches-Pereira, Series Editors for Springer Cities and Nature for their thoughtful comments on the manuscript preparation. Thank you also to Balaganesh Sukumar, Springer Nature for coordinating the production of this book.

Finally, a big thank you for the support of friends and family. To my mother for being a constant source of inspiration. To my father, brother, sister, and nieces for providing moral support. Also to Ken and Sylvia Maughan for their kindness. And to Lindsay for all her love and support throughout all stages of my research, for her infectious drive and determination, and especially her good humour. Thank you xx.

Contents

About the Author

Dr. Nicholas Ardill is an architect, lecturer and post-graduate researcher. He holds a Ph.D. in Architecture and Urbanism awarded by the University of Portsmouth and a Post-Graduate Diploma and Masters Degree in Architecture from the University of Brighton, where he qualified as an Architect in 2014. First entering practice in 1998, he has worked in Winchester and London for leading architectural studios before starting his own practice. His extensive experience and knowledge in architectural design and construction processes are complemented with a passionate interest for research into sustainable cities and green urbanism. Since 2014, he has served as a visiting lecturer in the School of Architecture at the University of Brighton. His current research focus is on socio-spatial processes and dynamics around urban change. He has authored papers in peer-reviewed journals on urban planning and sustainable development.

List of Figures

List of Tables

Part I
Framing the Research

Chapter 1
Social Innovation as a Collaborative Approach to Urban Food Spaces and Sustainable Development

Abstract This chapter introduces the concept of 'social innovation', and outlines what it is, who is doing it, and why it is significant to sustainable development. The central argument concerns how little attention has been paid to the process of social innovation in the production of urban spaces, especially public spaces for food growing. The chapter will provide background to the topic, including an overview and evaluation of existing research. It discusses the growing interest on social innovation as a solution for local to global problems, the people and spaces influencing innovation, and the ways in which contextual dynamics affect change processes. Urban food growing is proposed as social action to contribute to the challenges of contemporary cities such as public access to green space, participation in decision-making processes, and urban sustainability. How this novel spatial practice can meet needs, create social relationships, and form new collaborations is discussed. Finally, the research focus and questions are stated, and an overview of conceptual contributions is presented.

Keywords Social innovation · Urban food growing · Co-production of space · Urban green space · Socio-spatial dynamics · Social action · Sustainable development · Sustainable cities

This book investigates how social innovation develops in urban spaces. It presents exploratory research into the various phases in which people will interact to produce urban food spaces as a result of innovation.

Social innovation has multiple meaning though is understood generally as 'innovations that are social in both their ends and their means' (Mulgan 2019, p. 10), and it has become prevalent in political discussions. In a report by the Bureau of European Policy Advisers, social innovations are specifically defined as:

This chapter draws from previously published: Ardill, N. & Lemes de Oliveira, F. (2018) Social innovation in urban spaces, *International Journal of Urban Sustainable Development*, 10:3, 207–221, https://doi.org/10.1080/19463138.2018.1526177;

Ardill, N. & Lemes de Oliveira, F. (2021) Emerging Places of Social Innovation (POSI): a Conceptual Framework for Social innovation in Cities, *Transactions of the Association of European Schools of Planning,* 5:1, 55–70, https://doi.org/10.24306/TrAESOP.2021.01.005.

N. Ardill, *Growing Food in Cities*, Cities and Nature,
https://doi.org/10.1007/978-3-030-98475-5_1

> …new ideas (products, services and models) that simultaneously meet social needs (more effectively than alternatives) and create new social relationships or collaborations. They are innovations that are not only good for society but also enhance society's capacity to act (BEPA 2011, p. 9).

In this mutual approach, the process of social innovation in cities arises from in the interactions between individuals and groups of people who are engaged in meeting social goals. Collective action is taken by multiple stakeholders, including local inhabitants, professionals, and policy makers with a common interest in solving a social issue, thereby producing social capital and empowering beneficiaries (Ardill and Lemes de Oliveira 2018).

1.1 Social Innovation for Addressing Challenges to Sustainable Cities

In the context of urban austerity, combined with a growing interest in collaborative approaches to spatial development, societal challenges have brought a renewed attention to the potential roles of social innovation to sustainable development policy. Social innovation is positioned by the European Commission (2013a, b) as an important element in addressing grand challenges, such as poverty and social exclusion, the rising cost of healthcare and wellbeing, access to healthy food and food resilience, and climate change. Societal challenges are ever-increasingly experienced in urban areas as over half of the world's population now resides in cities—a trend projected to rise to 70% by 2050 (Newman and Jennings 2008; UNFPA 2016). Contemporary cities therefore have an imperative to find more effective and efficient solutions for tackling local to global problems.

Geels and Schot (2016) suggest that the field of innovation studies is well-placed to contribute to debates on urban transitions to tackle such challenges, but only when it considers the role of human agency in transforming built environments towards sustainable development. In particular, Mulgan (2006) points out how there have been few studies on the process dimension of social innovation that have investigated common patterns or combined learning, especially how it is 'designed, diffused and supported' (Caulier-Grice et al. 2012, p.33). Moulaert and Mehmood (2011) further argue that sector-specific research on the spatial characteristics of social innovation is needed to understand the dynamics in social and urban change.

The timing of my research is especially relevant as the 2020 global Covid-19 pandemic has reaffirmed the importance of public green space, urban food growing, and food security in cities; ensuring citizen access to fresh produce and the multiple health and social benefits generated. Furthermore, it has revealed large inequalities in access to green and natural space. According to the Office for National Statistics, one in eight UK households (12%) had no access to a private or shared garden during the 2020 lockdown while people from minority ethnic groups and low income households were most likely to have no access to a private garden (Office for National Statistics

Fig. 1.1 Urban food growing in public spaces, showing residents that it is possible to grow their own food. Stanford and Cleveland Community Garden, Brighton & Hove, UK (*Source* Author)

2020). Access to local green and natural spaces is significant as it is understood to benefit people's mental wellbeing, leading to reductions in depression, anxiety and obesity (The Wildlife Trusts 2020). The recent reliance on public open space has emphasised the need for more green spaces in urban areas, including community spaces for food growing (Fig. 1.1). While urban land resources are becoming tighter, many towns and cities still have unused and underused spaces, presenting an opportunity to transform these wasted spaces into productive spaces to meet the needs of local inhabitants.

In terms of wider socio-ecological value, urban spaces developed for food growing can support natural cooling and reduction of air pollution in cities, carbon sequestration, protection of biodiversity, and help people to lead more sustainable lifestyles (Artmann and Sartison 2018; Audate et al. 2018). As I will explain later, these productive spaces have the added potential to give local inhabitants more ownership of green areas by activating residual urban spaces, developing new meeting places and focal points for learning about the environment. In enhancing the identity of a city, these spaces can also promote food growing, fresh food in the diet, and health and wellbeing through contact to nature.

Social innovation spaces, scales and dynamics

Social innovation develops across various geographical scales. This is emphasised within two perspectives of innovation and its relationship to social change. The research on 'Transformative Social Innovation' models these scales as micro, meso, and macro (Haxeltine et al. 2013; Avelino et al. 2014). This global perspective of systemic change is based on the 'multi-level perspective' of sustainability transitions

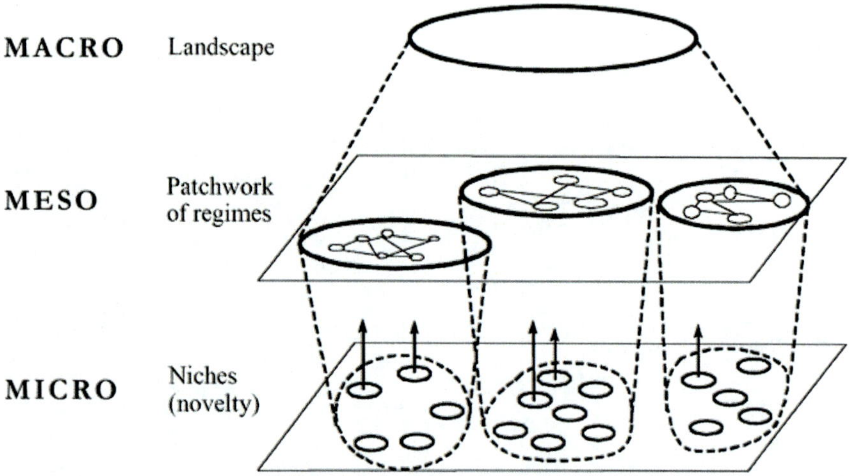

Fig. 1.2 Multi-level perspective of innovation (adapted from Geels 2002)

offered by Geels (2002). As shown in Fig. 1.2, these models propose that innovations develop in niche spaces (micro scale), cause changes to existing regimes of production and consumption (meso scale), and over time transform socio-technical landscapes (macro scale) (see, e.g. Rip and Kemp 1998; Geels 2002; Haxeltine et al. 2013).

The multi-level perspective implies interactions between multiple innovation participants, including individuals, organisations, and institutions from different economic sectors in an upwards direction of innovation across scales (Geels 2002). Whilst this perspective is useful for understanding longer-term societal change it has been much criticised (e.g. Berkhout et al. 2004; Seyfang and Smith 2007; Hargreaves et al. 2011). In particular, the model offered situates localised innovation activity within a broad model, neglecting the sources and contexts of this action.

Manzini (2015) provides an alternative perspective on innovation and scale in the research on 'Design for Social Innovation'. This approach focuses on a variety of participants; citizens, non-profit organisations, enterprises, and institutions all interacting at the local level with communities and neighbourhood users to collaboratively design sustainable solutions. It emphasises smallness, localism, design intermediation and 'co-creation' in emerging urban systems and environments.

While both perspectives offer understandings into how innovation develops, they investigate this activity at different scales. The urban spaces and outputs of social innovation, understood by Ardill and Lemes de Oliveira (2019, 2021) as 'Places of Social Innovation', are situated within these scales. They are settings in which place-based innovators collaboratively plan, design, and deliver urban spaces whilst also influencing larger national scale environmental and societal changes. Community food growing spaces as an output of socially innovative action are linked here to the strategies of local groups and institutions generating impetus behind spatial interventions. Especially, in the material, sociocultural and political structures of food growing initiatives, in addition to the shared distribution of public resources (e.g. land) and local user experiences at the urban scale. Moreover, these emerging productive green spaces concern innovation and experimentation in tailoring socio-ecological solutions to urban situations through participatory development processes.

'Agents of change': social innovation and sustainable urban development

I approach the study of social innovation in the production of urban spaces through particular 'agents of change', categorised in three levels: residents, non-governmental organisations, and local authorities (see Fig. 1.3). Agents of change refers to the various individuals, stakeholders and organisations who instigate innovation and approaches to participatory development across societies and cities. Existing studies have suggested that social innovation might be understood as a blend of public, private, and civil society participants coming together through an exchange of ideas and values. Baker and Mehmood (2015) propose that this interaction results 'in corresponding shifts in their roles and relationships as co-mingled agents of social change' (p. 10). However, as Howaldt et al. (2016, 2018) point out, it is only lately that hybrid arrangements of cross-sector agency have been investigated, and none have markedly examined social innovation processes in the production of urban spaces through a multilevel stakeholder analysis.

Existing research has emphasised the emergence of innovations within the public sector and from civil society in the urban environment. Adams and Arnkil (2013)

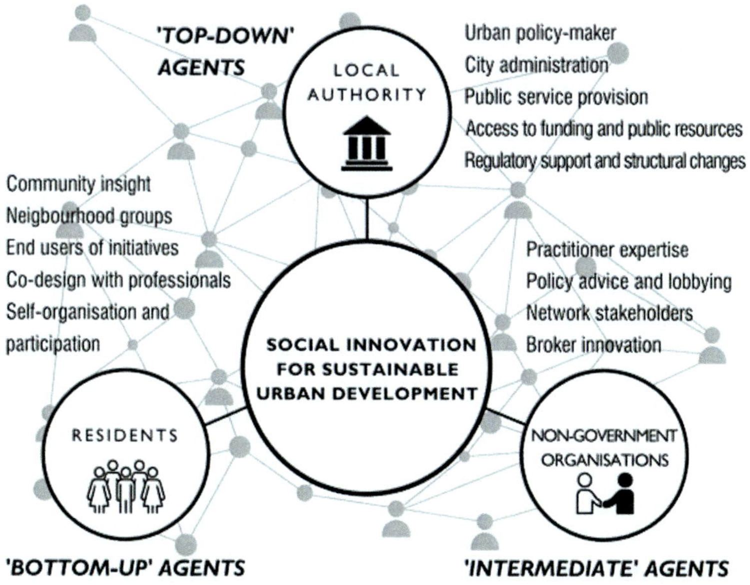

Fig. 1.3 Social innovation 'agents of change' for sustainable urban development (*Source* Author)

concluded that cities can become catalysts and innovation brokers in highlighting the role of city leadership, systems, and infrastructures as central to facilitating social innovation, especially to generate new exemplars of citizen participation. Manzini (2015), in addressing the role of 'creative communities', stresses how residents being more active within their local neighbourhood in undertaking collaborative action to regenerate the social fabric in cities may lead to empowerment, and consequently an urban system becoming more sustainable. Jegou and Bonneau (2015) suggest that engaging citizens more actively in producing collaborative public services through increasing user participation 'is not only possible, but is expected by citizens themselves' (p. 35). Furthermore, local authorities should create spaces for social innovation and collaboration with citizens to develop solutions to sustainable urban development. While Christmann et al. (2020) contend that supportive and locally influential participants in key positions are required from across sectors and at different scales to mutually engage in finding solutions. They conclude that a deeper understanding is required of the dynamic character of these interactions and their various complexities (supporting and constraining) during the process of social innovation, the various groups of participating stakeholders, and the shifting interactions and groupings among those concerned.

1.2 Food Growing in Cities as an Outcome of Social Action

I examine social innovation processes in the production of community food growing spaces in three spatial forms: food gardens, orchards, and productive urban landscapes extending across neighbourhoods (Fig. 1.4). Following Veen et al. (2016) community food growing spaces are understood both as an area of urban land, cultivated communally by a group of residents from the immediate neighbourhood or the wider city, and land in which urban inhabitants can participate in additional ways to gardening, but with a collective aspect. This communal aspect can be a shared responsibility for food growing and land maintenance, though can also involve collective governance structures (Knapp et al. 2016). Within and across different cities, these spaces integrate different characteristics and are adapted to the local context whilst involving different levels of cooperation between citizens, non-governmental organisations, and local authorities.

The concept of community food growing has been appropriated by various statutory and non-profit agencies as a socio-spatial intervention to help meet social needs and support urban regeneration, cohesion and health (Kingsley and Townsend 2006). Consequently, this social process is reshaping urban landscapes, experimenting with alternatives to capitalist formations of urban environment, and collaboratively producing public green spaces.

Community food growing spaces are now widely recognised as social innovations in themselves or as a product of social innovation process (see, e.g. Balian et al. 2016; Cunk et al. 2017; Spijker and Parra 2018; Tornaghi and Van Dyck 2015). These emerging urban spaces are proposed by Ardill and Lemes de Oliveira (2019)

Fig. 1.4 Urban food growing spaces developed through social innovation within Brighton & Hove, UK (*Source* Author; Additional photos, Brighton & Hove Food Partnership 2015)

to enable participation in a collective activity, whilst being publicly accessible, and produce social value in the transformation of urban spaces. Moreover, increasing public access to space is a significant condition in order to satisfy social needs and tackle societal challenges. For example, Mitchell (2003) draws attention to the increasing privatisation of public space and the exclusion of marginalised groups as a 'spatial injustice'. As Harvey (2009) observes, the issue of closure of urban space as an authoritarian act is problematic and there is a need to realise alternatives to capitalist means of spatial production. Furthermore, the relationship between enclosure, the privatisation of public space, and the future of cities is considered a struggle for a new 'urban commons', and this transformation of urban space is proposed as a collective, cultural and physical formation by and for its inhabitants (Harvey 2012).

As a common practice, urban gardening has a long history with spaces being cultivated to support subsistence. During the early-to-mid-twentieth century these sites multiplied in cities, especially in response to specific events and where crises and emergencies arose through eras of dramatic social change, such as the development of 'Victory Gardens' during World War I and World War II. Pudup (2008) suggests that contemporary experimentation with community food growing to support a more socio-political agenda can be traced back to the emerging environmental debates and social movements from the 1970s onwards, including the 'community garden movement' in the reclaiming and reusing of urban land. Peck and Tickell (2002) point out that amid the increasing 'neoliberalisation' of public space in the 1980s, attributed to state budget cuts and the reduction of state interventions, these green spaces became a form of countermovement against urban land commodification by market forces and were seen as a source of collective empowerment for local inhabitants. There has since been a resurgence of Community food growing spaces as a 'contested space' in which disadvantaged communities attempt to maintain their presence in the built environment amid ongoing urban renewal and gentrification processes. For example, I analyse in one of my case studies how local residents came together to form a cooperative and develop a community hub with interlinked food growing spaces, helping to protect a neighbourhood social space from commercial redevelopment (see Chap. 6 'The Bevy Edible Garden', Brighton & Hove). Consequently, over time and across varying urban settings these types of productive green spaces have evolved, leading to multiple meanings, with the spaces varying in design, shape, and size.

I agree with Firth et al. (2011) in differentiating between municipal allotment gardens and community food growing spaces. In the UK, contemporary allotment gardening involves individual members hiring a plot of land for an annual subscription from the local authority within an enclosed space with controlled access. Whereas community food growing is more 'public' in terms of spatial ownership, access and degree of democratic control. The community food growing spaces I studied across the cities of Brighton & Hove and Portsmouth, UK can be characterised as common spaces that bring people together through collective action. These shared spaces typically act as hubs for the community to foster social interaction and social networks, and aside from providing space within urbanised environments to connect with nature to support wellbeing, can also support capacity building through education, skills development and training. Moreover, they can be understood as 'third places'; a space outside the home and work that creates possibilities for people to meet and interact in a public setting (see, e.g. Oldenburg 2001). Nonetheless, as Pudup (2008) points out, the use of the term 'community' in relation to urban food growing can present multiple understandings. From my observations, it is not always distinct whether these spaces are governed by the community for the community, or just situated within certain communities.

Firth et al. (2011) proposes two categories of community gardens; place-based and interest-based. Place-based forms are embedded in the local community. They are typically bottom-up driven and developed by and for residents within close neighbourhood proximity who strongly identify with that place. Whereas, interest-based are initiated and managed by individuals or groups from outside the local community who are bound together through common interest. As I demonstrate later, these external agents may span across diverse communities and can involve professionals from governing institutions and/or non-profit organisations within spatial governance. Firth et al. (2011) acknowledge that the two categories may not always map neatly on to one space as they may involve hybrid stakeholder arrangements. Both, however, share an important characteristic in relation to the process of social innovation; participants attempt to collaboratively develop urban spaces by fostering social relations, whether in a horizontal or vertical direction (e.g. with community groups or local authorities).

In summary, social innovation in the development of urban food spaces is of interest as a participatory concept that can meet human needs, create social relationships, and form new collaborations. However, as a process this innovative approach does not necessarily conclude with the realisation of a food growing space. Like cities, this social action needs to continuously evolve in order to influence social structures and systems with regard to participation in decision-making processes, social inclusion, and sustainable development.

1.3 Research Focus and Questions

The research outlined in the previous sections of this chapter argues for the promise of collective action to influence social and urban changes towards realising sustainable development. However, these studies and publications raise questions of the conceptions of social change in existing theories. In particular, social innovation theories do not sufficiently pay attention to the sequences in which social action produces urban changes and the environmental forces that develop or diminish this action. Currently, there is a lack of understanding on the processes involved in 'spatial production', understood as the transformation of physical space and spatial relations.

I therefore concur with Brandsen et al. (2016) and Howaldt et al. (2016, 2018) in that there are outstanding questions concerning the social innovation agency of cross-sector collaborations within built environments to influence processes of urban change. I will explore these omissions in this book. In particular, my central aim is to advance the understanding of social innovation by formulating a process model of spatial production. In so doing, I examine the actions of urban stakeholders to produce food growing spaces and the contextual influences that affect this innovation activity. This book is directed by my underlying research question:

> How are food growing spaces in cities co-produced between urban stakeholders at three different organisational levels of social innovation, characterised as residents, non-governmental organisations, and local authorities?

My research will build on understandings from studies on social innovation in urban spaces in combination with insights on social innovation process to formulate a conceptual framework. In order to test and develop this framework, two specific questions are posed:

> 1): Are there any observable patterns to the key social innovation processes undertaken in the production of urban food spaces and, if present, what accounts for these?

> 2): How do contextual forces across spatial scales influence the social innovation actions of urban stakeholders in the production of food growing spaces?

My research advances the present understanding of the process of social innovation in urban change. Interest to-date has been directed either towards the scaling up of successful social innovations to transform societies and realise systems change for sustainability (see, e.g. Mulgan 2019), or increasing the participation of socially excluded groups in urban decision-making processes (see, e.g. Moulaert and MacCallum 2019). In seeking patterns to social innovation in spatial production, I determine an 'ideal type' sequence of innovation, whilst identifying contextual forces across geographical scales affecting stakeholder action in developing urban food spaces.

I further the recognition of what Howaldt et al. (2018) propose as the combination of 'social innovation actors'; those various urban stakeholders who are instigating social innovation processes and urban outcomes. In so doing, I contribute to ongoing debates on the meaning of social innovation, including radical versus complementary perspectives, and its relationship to structural and systemic changes.

References

Adams E, Arnkil R (2013) Supporting urban youth though social innovation: stronger together Urbact II. URBACT, Brussels

Ardill N, Lemes de Oliveira F (2018) Social innovation in urban spaces. Int J Urban Sustain Dev 10(3):207–221. https://doi.org/10.1080/19463138.2018.1526177

Ardill N, Lemes de Oliveira F (2019) Emerging places of social innovation (POSI). The co-production of space between multilevel stakeholders. In: "Planning for transition", Annual congress for the Association of European Schools of Planning (AESOP), Aesop 2019 Conference, Venice, 9–13 July, AESOP, Venice, pp 2303–2319

Ardill N, Lemes de Oliveira F (2021) Emerging places of social innovation (POSI): a conceptual framework for social innovation in cities. Trans Assoc Eur Sch Plan 5(1):55–70. https://doi.org/10.24306/TrAESOP.2021.01.005

Artmann M, Sartison K (2018) The role of urban agriculture as a nature-based solution: a review for developing a systemic assessment framework. Sustainability 10(6):1937

Audate PP, Fernandez MA, Cloutier G, Lebel A (2018) Impacts of urban agriculture on the determinants of health: scoping review protocol. J Med Internet Res 20(3):1–14

Avelino F, Wittmayer J, Haxeltine A et al (2014) Game changers and transformative social innovation. TRANSIT, Brussels

Baker S, Mehmood A (2015) Social innovation and the governance of sustainable places. Local Environ Int J Justice Sustain 20(3):321–334

Balian E, Berhault A, Eggermont H et al (2016) Social innovation and nature-based solutions. EKLIPSE, Brussels

BEPA (2011) Empowering people, driving change social innovation in the European Union. European Commission, Brussels

Berkhout F, Smith A, Stirling A (2004) Socio-technological regimes and transition contexts. In: Elzen B, Geels F, Green K (eds) System innovation and the transition to sustainability: theory, evidence and policy. Edward Elgar Publishing Limited, Cheltenham, pp 48–75

Brandsen T, Cattacin S, Evers A, Zimmer A (eds) (2016) Social innovations in the urban context. Springer, London

Brighton & Hove Food Partnership (2015) Brighton & Hove sustainable food cities award 2015. Brighton & Hove Food Partnership, Brighton & Hove

Caulier-Grice J, Davies A, Patrick R, Norman W (2012) Defining social innovation. European Commission, DG Research, Brussels

Christmann GB, Ibert O, Jessen J, Walther U-J (2020) Innovations in spatial planning as a social process—phases, actors, conflicts. Eur Plan Stud 28(3):496–520

Cunk K, Straus M, Zamfira R (2017) Approaching urban agriculture as a social innovation. AgriGo4Cities, Koper

European Commission E (2013a) Horizon 2020 EU research and innovation: tackling societal challenges. European Commission, Brussels
European Commission E (2013b) Guide to social innovation. European Commission, Brussels
Firth C, Maye D, Pearson D (2011) Developing "community" in community gardens. Local Environ 16(6):555–568
Geels FW (2002) Technological transitions as evolutionary reconfiguration processes: a multi-level perspective and a case-study. Res Policy 311257–1274
Geels FW, Schot J (2016) Towards a new innovation theory for grand societal challenges. SPRU anniversary conference. University of Sussex, Brighton & Hove, pp 1–37
Hargreaves T, Haxeltine A, Longhurst N, Seyfang G (2011) Sustainability transitions from the bottom-up: civil society, the multi-level perspective and practice theory. CSERGE, Norwich
Harvey D (2009) Social justice and the city, revised edn. University of Georgia Press, Athens
Harvey D (2012) Rebel cities: from the right to the city to the urban revolution. Verso Books, London
Haxeltine A, Avelino F, Wittmayer J et al (2013) Transformative social innovation. In: Social frontiers. Nesta, London
Howaldt J, Kaletka C, Schröder A (2016) Social entrepreneurs: important actors within an ecosystem of social innovation. Eur Public Soc Innov Rev 1(2):95–110
Howaldt J, Kaletka C, Schröder A, Zirngiebl M (eds) (2018) Atlas of social innovation. TU Dortmund University, Dortmund
Jegou F, Bonneau M (2015) Social innovation in cities, URBACT II. URBACT, Saint Denis
Kingsley J, Townsend M (2006) "Dig in" to social capital: community gardens as mechanisms for growing urban social connectedness. Urban Policy Res 24(4):525–537
Knapp L, Veen E, Renting H et al (2016) Vulnerability analysis of urban agriculture projects: a case study of community and entrepreneurial gardens in the Netherlands and Switzerland. Urban Agric Reg Food Syst 1(1):1–13
Manzini E (2015) Design, when everybody designs: an introduction to design for social innovation. MIT Press, Cambridge, Massachusetts
Mitchell D (2003) The right to the city: social justice and the fight for public space. Guilford Press, New York
Moulaert F, MacCallum D (2019) Advanced introduction to social innovation. Edward Elgar Publishing, Cheltenham
Moulaert F, Mehmood A (2011) Spaces of social innovation. In: Pike A, Rodriguez-Pose A, Tomaney J (eds) Handbook of local and regional development. Routledge, Abingdon, pp 212–225
Mulgan G (2006) The Process of Social Innovation. Innov Technol Gov Glob 1(2):145–162
Mulgan G (2019) Social innovation: how societies find the power to change. Policy Press, Bristol
Newman P, Jennings I (2008) Cities as sustainable ecosystems: principles and practices. Island Press, London
Office for National Statistics (2020) One in eight British households has no garden. https://www.ons.gov.uk/economy/environmentalaccounts/articles/oneineightbritishhouseholdshasnogarden/2020-05-14. Accessed 14 May 2020
Oldenburg R (ed) (2001) Celebrating the third place: inspiring stories about the "great good places" at the heart of our communities. Da Capo Press, Chicago
Peck J, Tickell A (2002) Neoliberalizing space. Antipode 34(3):380–404
Pudup MB (2008) It takes a garden: cultivating citizen-subjects in organized garden projects. Geoforum 39(3):1228–1240
Rip A, Kemp R (1998) Technological change. In: Rayner S, Malone E (eds) Human choices and climate change. Battelle, Columbus, pp 327–399
Seyfang G, Smith A (2007) Grassroots innovations for sustainable development: towards a new research and policy agenda. Env Polit 16(4):584–603
Spijker SN, Parra C (2018) Knitting green spaces with the threads of social innovation in Groningen and London. J Environ Plan Manag 61(5–6):1011–1032

The Wildlife Trusts (2020) Nature for wellbeing. https://www.wildlifetrusts.org/nature-health-and-wild-wellbeing. Accessed 19 Apr 2020

Tornaghi C, Van Dyck B (2015) Research-informed gardening activism: steering the public food and land agenda. Local Environ 20(10):1247–1264

UNFPA (2016) Urbanization. www.unfpa.org/urbanization. Accessed 24 Oct 2016

Veen EJ, Bock BB, Van den Berg W et al (2016) Community gardening and social cohesion: different designs, different motivations. Local Environ 21(10):1271–1287

Chapter 2
The Study Sites and Methods

Abstract This chapter explains the methods adopted and the theory supporting them in relation to the aims and research questions of the study. It outlines the process approach to this research, justifying the perspective due to seeking explanations in patterns, dynamics, and their relationship with time. Additionally, it describes the research design and case study strategy, including data collection, the selection of research sites, and social innovation agents under study. The chapter explains why six case studies of emerging urban food spaces were examined across two distinct cities, with different catalysts and obstacles, and with particular opportunities and needs. These cities are Brighton & Hove and Portsmouth, UK. Until 2001, The City of Brighton & Hove was two separate towns: Hove and Brighton. This vibrant seaside resort has an extended history as a centre of socio-ecological activism, observed across several spheres of sustainable development, especially food growing initiatives. Portsmouth, granted city status in 1926 and a significant naval port for centuries, does not share an equivalent history of socially innovative action in sustainable practices nor experimentation in alternative development, and for a city as populated as Portsmouth there is limited activity around urban food growing.

Keywords Process research · Innovation patterns · Social innovation · Co-production of space · Urban food growing · Sustainable development · Brighton & Hove · Portsmouth

2.1 Process Research Approach

I employ a process research approach to investigate how social innovation processes occur and evolve in the production of urban food spaces. This method places human agency and the context of action foremost when accounting for processes of change. Moreover, the interactions between agency and context occur across time in a cumulative manner and the context both shapes and is shaped by this social action (Ardill

---

This chapter draws from previously published: Ardill, N. & Lemes de Oliveira, F. (2021) Emerging Places of Social Innovation (POSI): a Conceptual Framework for Social innovation in Cities, *Transactions of the Association of European Schools of Planning*, 5:1, 55–70. https://doi.org/10.24306/TrAESOP.2021.01.005

N. Ardill, *Growing Food in Cities*, Cities and Nature,
https://doi.org/10.1007/978-3-030-98475-5_2

and Lemes de Oliveira 2018). In this sense, urban stakeholders are understood as both products and producers of their environment and social system (see Chap. 1).

A process approach permits enquiry into several aspects for analysis and is therefore well suited to this study. By examining the process of social innovation and contextual influences across multiple scales and organisational levels, I contribute new insight on this social action in spatial production by investigating the topic from

Table 2.1 Characterisation of the urban stakeholders who instigate social innovation processes in the co-production of food growing spaces

Type of social innovation	Typical governance structures	Examples and indicators of social action (with reference to innovation literature)
Urban stakeholder: residents		
Community based	Co-operative. Management committee led by residents' group co-ordinator	Participatory decision-making processes at the neighbourhood scale (micro) through the involvement of local citizens (Moulaert et al. 2010); neighbourhood driven agency embedded in the responses of indigenous communities empowering themselves to act in their interests (Seyfang 2004); activity ranges from self-help to radical engagement to develop processes of social change (Pestoff 2014)
Urban stakeholder: non-governmental organisations		
Practitioner (specialist) based	Board of directors and management committee led by the organisation co-ordinator	Civil society organisations aiming to diffuse innovative ideas, activities, and practices. Mediation activities occur between levels of urban stakeholders and spatial scales (micro to meso) (TEPSIE 2014); agency includes providing advice, training, network building, project initiation, management and coordination (Hodson and Marvin 2010); they encourage local to national government, through the agency of lobbying, to pursue new solutions that lead to change (Seyfang et al. 2010)
Urban stakeholder: local authorities (government institutions)		

(continued)

Table 2.1 (continued)

Type of social innovation	Typical governance structures	Examples and indicators of social action (with reference to innovation literature)
Policy-based (public sector)	The full council led by the leader of the council; committees include policy and regulatory, policy and resources, and overview and scrutiny	Policy decisions made by local authority governmental officials at the city scale (meso) and between local and national scale policy (macro); access to levers including funding and support to accelerate innovative activity. Can provide structural changes, steering and regulatory support (Howaldt et al. 2018)

Source Author

a multidimensional perspective. Process analysis is presented here as case narratives which describe the sequences of events through which change occurs. The innovativeness of this application concerns outcomes in social and urban change being explained in terms of methods and patterns with the sequencing and time ordering of processes being important to their construction. Furthermore, the ability to generalise explanations is reliant on the extent to which they can clarify various patterns of urban development. In order to evaluate the building of theory in my study, process research measures were used. Following the advice of Pentland (1999), Van de Ven and Poole (2005), and Langley et al. (2013), I use research measures of generality together with accuracy, in the extent that the resultant theory accurately depicts empirical evidence, and simplicity, regarding the components to a theory, to support validity.

The strategy of a longitudinal and comparative case study complements the research focus. It enables the tracking of innovation processes across several scales and chronologies, enhances pattern recognition, and enables the identification of relationships between key processes. Moreover, the multiple case study design supports the cross-case comparison and triangulation of results.

I examined six case studies of emerging food growing spaces across two UK cities: Brighton & Hove and Portsmouth. Three cases were selected from each city with one case for each urban stakeholder level of social innovation: residents; non-governmental organisations; and local government institutions (see Table 2.1). This strategy builds upon the recommendations of Pettigrew (1990) to use 'polar types' and 'extremes' (e.g. conditions within cities), and Flyvbjerg's (2006) suggestion of 'maximum variation' (e.g. type of urban stakeholder) to obtain information about the significance of various circumstances for case process and outcomes. Furthermore, two-pairs of three cases provides a practicable method to support basing wider conclusions in relation to the requirement for depth of process data and external validity (Van de Ven and Poole 2004).

2.2 Research Case Studies

The six cases under detailed study occurred between 2008 to 2018 and are located within two urban areas of Brighton & Hove and Portsmouth on the South East coast of the UK. I will now introduce the two cites chosen and provide a summary of the selected case studies.

Brighton & Hove: is a city of approximately 274,000 residents covering a compact geographic territory, constrained by the sea and the South Downs National Park (Fig. 2.1). It has an extended history as a centre of socio-ecological activism in relation to alternative development pathways (Devereux 2014). Action to tackle contemporary challenges to sustainable development is observed across several spheres of sustainability, including energy, transport, waste, and especially food growing initiatives (ARTS 2016). For example, the Brighton & Hove Food Partnership was established in 2003 as an independent not-for-profit advocacy organisation to create a more sustainable local food system (Ardill and Lemes de Oliveira 2021). Their innovation in adopting a strategic partnership approach to integrating food work across a range of actions in the city has involved collaborating with the local authority to adopt the first urban food strategy in the UK in 2006. The organisation instigated the development of 54 new community food growing spaces between 2009 to 2013 through their 'Harvest Brighton & Hove' urban development programme, tripling the number of food growing projects from 25 to 79, with several projects in new spaces including parks, estates and railway stations (White et al. 2013). Through lobbying,

Fig. 2.1 View of the Brighton & Hove seafront (*Source* Knight 2012, CC BY-SA 2.0)

in 2011 the city became the first in the UK to introduce planning guidance recommending spaces for food growing be integrated within new developments. Brighton & Hove was also the first city in 2015 to be awarded what was then the highest 'Silver Sustainable Food Cities Award', subsequently 'Gold' in 2020, another first, owing to its widespread community of 'trail-blazers' and 'innovators' and because the city is 'leading the way of one of the fastest growing social movements in the UK today' (Clarke 2015, p. 1).

Portsmouth: is the UK's only 'island city' of approximately 205,000 residents and occupying only 40 square kilometres, it is the most densely populated city in the UK after London (Food Matters 2015). The city also faces geographical challenges to sustainable development given its location, surrounded by the sea on three sides and hills to the north, making available land for food growing scarce (Fig. 2.2). The importance of green space is particularly relevant due to the densely populated urban area. Land is valued, and it must accommodate a diversity of uses and variety of urban needs. In contrast to Brighton & Hove, the city does not possess an equivalent history of social innovation in sustainable practices nor experimentation in alternative development, and for a city as populated as Portsmouth there is limited activity around community food growing (Food Matters 2015). Attempts to diffuse the practice across the city have been met with mixed successes. For example, between 2009 to 2011, 11 community growing spaces were developed in deprived areas through a top-down urban health initiative of the local authority known as 'Healthy Pompey'. Three of which subsequently closed not long after opening. In 2015, Food Portsmouth was launched as an umbrella organisation to unite sectors within the city to support and promote a local and sustainable food system, including the development of

Fig. 2.2 Aerial photograph of Portsmouth on Portsea Island (*Source* Ministry of Defence 2005, OGL v1.0)

urban spaces to grow food. The organisation developed a community food growing space and provided support to grassroots groups. However, it has lacked the agency and resources of the equivalent Brighton & Hove organisation to increase urban development. Furthermore, initial plans made for introducing planning guidance, modelled on the experiences of Brighton & Hove, to integrate food growing into local development policy stalled following a draft proposal being submitted to the City Council for public consultation in 2015. In July 2017, there were 15 community food growing spaces in Portsmouth, in contrast to 82 such spaces in Brighton & Hove.

The cases under study from Brighton & Hove are shown geographically in Fig. 2.3 and are summarised as follows:

The Bevy Edible Garden: A residents' co-operative in Brighton & Hove transformed the site of a former public house into a community owned hub and social enterprise; comprising several productive green spaces interlinked with a community cafe and training kitchen (Fig. 2.4).

Racehill Community Orchard: A Brighton & Hove-based food advocacy organisation partnered with permaculture specialists to develop the largest community orchard within the city at 1.30 ha. As an exemplar project, it served as a template for 'Harvest Brighton & Hove', a development programme that produced over 50 community food growing spaces citywide (Fig. 2.5).

The Keep Community Orchard: This Brighton & Hove City Council development was integrated with an archive centre for East Sussex. As a precedent for urban planning, the aim was to test innovative approaches in helping the city find ways to transition towards sustainable development. The development influenced the adoption of planning policy to support urban food growing (Fig. 2.6).

The cases under study from Portsmouth are shown geographically in Fig. 2.7 and are summarised as follows:

Southsea Greenhouse: This resident led development in Portsmouth started at a seafront concession before transforming wasteland within a park into a community run social enterprise comprising productive green spaces. Residents formed a co-operative and sold shares in the community owned venture (Fig. 2.8).

Charles Dickens Orchard Trail: A Portsmouth-based environmental organisation developed a linear fruit orchard as a 'productive urban landscape' and heritage walking trail to support food growing practices and connect deprived neighbourhoods within a highly urbanised centre. A co-design approach with the local authority assisted its spatial planning and configuration (Fig. 2.9).

Stacey Community Orchard: This Portsmouth City Council initiative transformed an area surrounding a community centre into a demonstration space for food growing in small urban spaces. Central Government funding from a 'Healthy Towns' programme was used to trial innovative ways of tackling obesity by changing resident behaviour to live healthier lives, leading to the local development of 11 community food growing spaces (Fig. 2.10).

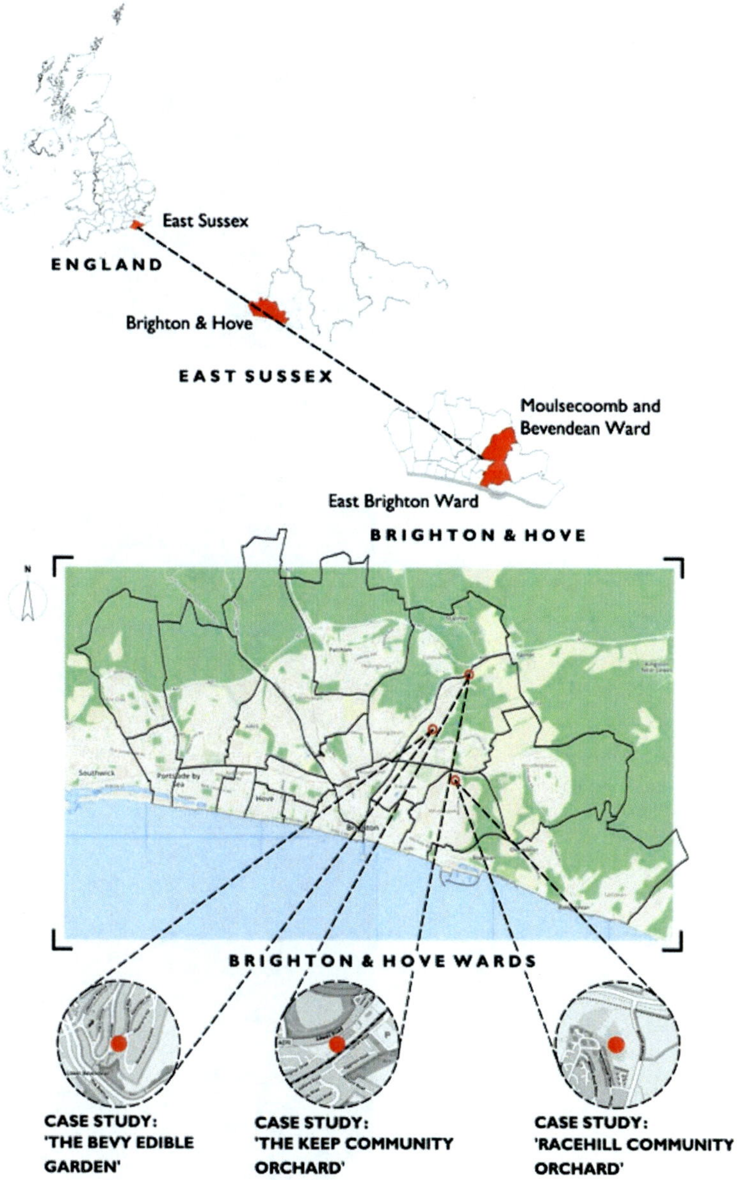

Fig. 2.3 Brighton & Hove case studies location (*Source* Author)

Fig. 2.4 The Bevy Edible Garden, Brighton & Hove (*Source* The Bevendean Cooperative 2017)

Fig. 2.5 Racehill Community Orchard, Brighton & Hove (*Source* Brighton & Hove Food Partnership 2015)

Fig. 2.6 The Keep Community Orchard, Brighton & Hove (*Source* Iliffe 2014)

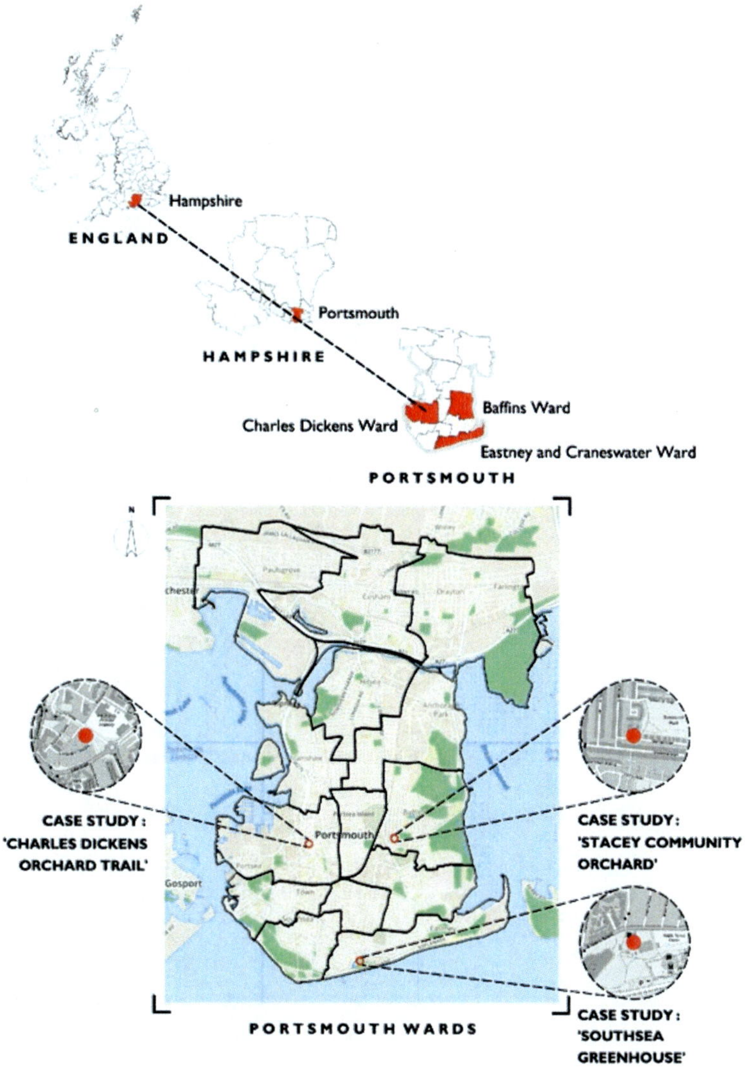

Fig. 2.7 Portsmouth case studies location (*Source* Author)

Fig. 2.8 Southsea Greenhouse, Portsmouth (*Source* Southsea Greenhouse Cooperative 2016)

Fig. 2.9 Charles Dickens Orchard Trail, Portsmouth (*Source* Bell 2019)

Fig. 2.10 Stacey Community Orchard, Portsmouth (*Source* Jones 2019)

2.3 Approach to Data Collection and Analysis

The data collected in this study was based on multiple information sources so that I could obtain a triangulation of data through analytical stages and to support theory-building. In following the advice of Flyvbjerg (2006), I employed a selective 'information-oriented' approach to gathering case data. The data sources comprised of semi-structured interviews with key informants within each case; content analysis of archival documents; and participant observation of innovation activities and environments.

I interviewed three main categories of urban stakeholders: residents steering neighbourhood initiatives, project leads of local non-governmental organisations, and local authority officers influencing policy. These participants were consulted on the basis of their knowledge of case activities and roles within cases, characteristically that of a project lead or key stakeholder in the planning, design, and delivery of an urban food growing space. This strategy meant that I interviewed a selective number of participants, though in seeking rich qualitative data, it meant that deeper, more probing interviews with expert witnesses were conducted to cover multiple perspectives central to case development.

A two-phase interviewing process progressed from being exploratory to more probing and structured. The interest here was in helping to establish development periods, decision-making processes, and confirm sequences of case events. Overall 17 key-informants were interviewed in-depth and face-to-face. Significantly, nine had been involved with the development of a minimum of two cases in their respective urban systems and therefore these interviewees were strategically selected to enable multiple case discussions. Furthermore, due to interviewing participants with a common interest within compact urban areas, it meant that most participants had previously interacted across initiatives and case events and therefore could provide added insight. In addition to these interviews, I held recurring discussions with case participants and other prominent urban stakeholders which occurred in an impromptu fashion. This ongoing dialogue further illuminated case events, relationships, and contextual influences.

My subsequent approach to analysing the qualitative data gathered used process-based methods that were complemented by a thematic mode of enquiry. I applied a conceptual framework developed through combining research on social innovation in urban spaces (see Chaps. 3 and 4) with process measures and a case narrative methodology derived from Pentland (1999). Additionally, I used a data-driven inductive approach to identify emerging themes in my case studies that was based on Braun and Clarke (2006). The process of interpreting and translating empirical data is presented below in order for clarity of explanation, though was in practice recurring to support investigation. The six stages of analysis are summarised as:

1. **Case narratives**: rich text narratives are constructed by outlining chronological sequences of events and the use of plot structures to support narration and

handle data in an accessible manner with each case sequence produced from data gathering;
2. **Data coding**: integrates preconceived theory driven codes directed by the conceptual framework with emerging data driven codes;
3. **Within case analysis of key social innovation processes**: a study protocol of common questions developed from the conceptual framework is examined within each case narrative. This approach enables a concentrated assessment and comparison of key innovation processes;
4. **Visual narrative mapping of individual cases**: positions case events and activities across space and time. Mapping offers concise conclusions to individual case narratives whilst preparing the ground for cross-case analysis and pattern searching in innovation activity across the cases;
5. **Pattern recognition across cases**: identifies and explains patterns of key social innovation processes in the co-production of urban food growing spaces across the cases by comparing case sequences to advance the conceptual framework;
6. **Analysis of multiscale contextual forces**: classifies multiscale influences on innovation processes at three contextual scales: neighbourhood, urban system, and national system, and incorporates observations from the case studies and insights from innovation research, to position and separate socio-spatial dynamics into different groupings (Table 2.2).

Table 2.2 Contextual influences on social innovation processes

Contextual scales	Description	Examples and indicators
Neighbourhood (Micro scale)	The community in close geographical proximity to initiatives, including their end-users	Personal and group activities or practices. Includes interests and experiences in spatial production. User demand
Urban system (Meso scale)	The configuration of stakeholders and organisations across multiple sectors; local government, private, civil society and academic institutions that may influence innovation	Local government policy. City infrastructure and resources. City marketing and positioning
National system (Macro scale)	Composed from the conditions, events, and external factors that influence the trajectory of innovation	National government policy. Funding bodies. Organisations and networks

Source Author

References

Ardill N, Lemes de Oliveira F (2018) Social innovation in urban spaces. Int J Urban Sustain Dev 10(3):207–221. https://doi.org/10.1080/19463138.2018.1526177

Ardill N, Lemes de Oliveira F (2021) Emerging places of social innovation (POSI): a conceptual framework for social innovation in cities. Trans Assoc Eur Sch Plan 5(1):55–70. https://doi.org/10.24306/TrAESOP.2021.01.005

ARTS (2016) Roadmap for a Sustainable City Brighton and Hove. University of Sussex, Brighton

Bell P (2019) Cornwallis Crescent Community Orchard volunteers. https://www.facebook.com/pg/Cornwallis-Crescent-Community-Orchard-Volunteers-1009344429141432/photos/?ref=page_internal. Accessed 14 Sep 2019

Braun V, Clarke V (2006) Using thematic analysis in psychology. Qual Res Psychol 3(2):77–101

Brighton & Hove Food Partnership (2015) Racehill Community Orchard Report August 2015. Brighton & Hove Food Partnership, Brighton & Hove

Clarke C (2015) Brighton & Hove win top sustainable food award. https://bhfood.org.uk/brighton-hove-win-top-sustainable-food-award/. Accessed 19 Dec 2018

Devereux C (2014) Food and planning in Brighton & Hove. Food Matters, Brighton & Hove

Flyvbjerg B (2006) Five misunderstandings about case-study research. Qual Inq 12(2):219–245

Food Matters (2015) A review of the food system in Portsmouth. Food Matters, Brighton & Hove

Hodson M, Marvin S (2010) Can cities shape socio-technical transitions and how would we know if they were? Res Policy 39(4):477–485

Howaldt J, Kaletka C, Schröder A, Zirngiebl M (eds) (2018) Atlas of social innovation. TU Dortmund University, Dortmund

Iliffe F (2014) Food growing planning advice note sustainable food cities. Brighton & Hove City Council, Brighton & Hove

Jones T (2019) The Stacey Community Centre. https://www.facebook.com/pg/TheStaceyCommunityCentre/photos/?tab=albums. Accessed 1 Sep 2019

Knight G (2012) Photos taken in Brighton on 20th June 2012. https://www.flickr.com/photos/8176740@N05/7563631490. Accessed 27 Nov 2020

Langley A, Smallman C, Tsoukas H, Van De Ven AH (2013) Process studies of change in organization and management: Unveiling temporality, activity, and flow. Acad Manag J 56(1):1–13

Ministry of Defence (2005) An aerial photograph of Portsmouth Dockyard. https://www.defenceimagery.mod.uk/fotoweb. Accessed 27 Nov 2020

Moulaert F, Martinelli F, Swyngedouw E, Gonzalez S (eds) (2010) Can Neighbourhoods save the city? Community development and social innovation. Routledge, Abingdon

Pentland BT (1999) Building process theory with narrative: from description to explanation. Acad Manag Rev 24(4):711–724

Pestoff V (2014) Collective action and the sustainability of co-production. Public Manag Rev 16(3):383–401

Pettigrew AM (1990) Longitudinal field research on change: theory and practice. Organ Sci 1(3):267–292

Seyfang G (2004) Time banks: rewarding community self-help in the inner city? Community Dev J 39(1):62–71

Seyfang G, Haxeltine A, Hargreaves T, Longhurst N (2010) Energy and communities in transition—towards a new research agenda on agency and civil society in sustainability transitions. University of East Anglia, Norwich, CSERGE

Southsea Greenhouse Cooperative (2016) Southsea Greenhouse. https://www.facebook.com/pg/SouthseaGreenhouse/photos/?tab=album&album_id=131268423599780&ref=page_internal. Accessed 8 Aug 2016

TEPSIE (2014) Social innovation theory and research: a summary of the findings from TEPSIE. European Commission, DG Research, Brussels

The Bevendean Cooperative (2017) The Bevendean Cooperative. https://www.facebook.com/pg/TheBevy/photos/?ref=page_internal. Accessed 1 Nov 2017

Van de Ven A, Poole M (2005) Alternative approaches for studying organizational change. Organ Stud 26(9):1377–1404

Van de Ven A, Poole M (eds) (2004) Handbook of organizational change and innovation. Oxford University Press, New York

White R, Devereux C, Borrill V, Crocker J (2013) A Growing Community Harvest Evaluation report 2009–2013. Brighton & Hove Food Partnership, Brighton & Hove

Part II
Social Innovation in Urban Spaces: A Conceptual Framework

Chapter 3
Social Innovation for Sustainable Development

Abstract This chapter examines present understanding on social innovation for sustainable development in order to integrate findings on the spatial characteristics of social innovation, co-operative processes, and environmental dynamics. It breaks down the social innovation definitions proposed, identifying a split between two types of principal meanings, understood as radical versus complementary. The central characteristics of social innovation are established, concerning the development of social relations, production of social value, and its spatial scales. The chapter moves then to discuss the existing knowledge on social innovation as a strategy to promote sustainable development. Reflections are given on existing frameworks of social and urban change, including the relationship with human agency, space and time. The findings suggest that in general the collaboration of local inhabitants as end users in place-based development are central to this process of change and that process itself is as significant as the outcome.

Keywords Social innovation · Urban space · Sustainable cities · Sustainable development · Societal challenges · Social innovation process

3.1 Social Innovation: Definitions, Characteristics, and Scales

Debates on social innovation have tended to intensify in periods of political and social instability. In the nineteenth century, social innovation was linked in a derogatory manner to revolutionary aspects of socialism, social reform and the social economy. During the 1930s it was associated with the rise of the welfare state, and in the 1960s with new social movements for emancipation and democratisation. From the end of the twentieth century to the 2000s it was given new impetus and positive meaning, first in the context of local development for deindustrialising cities, and second by a renewed interest in the social and solidarity economy for welfare provision after the financial crises of 2008. Yet, there is no consensus on the definition of social innovation today. A conclusion drawn to by, among others, Mulgan et al. (2007), Pol

---

This chapter draws from previously published: Ardill, N. & Lemes de Oliveira, F. (2018) Social innovation in urban spaces, *International Journal of Urban Sustainable Development*, 10:3, 207–221. https://doi.org/10.1080/19463138.2018.1526177

N. Ardill, *Growing Food in Cities*, Cities and Nature,
https://doi.org/10.1007/978-3-030-98475-5_3

and Ville (2009), Rüede and Lurtz (2012), The Young Foundation (2012), Moulaert et al. (2013) and Benneworth and Cunha (2015).

Taken in isolation, the definitions of 'social' and 'innovation' seem no less contentious. Social is commonly understood as the production of social value to the public, or society as a whole, in contrast to privately accrued profit as proposed by Phills et al. (2008), Murray et al. (2010), BEPA (2011) and TEPSIE (2014). As for social value, it is often presented as addressing social needs or challenges, or improving capacity to act, wellbeing and quality of life (see e.g. BEPA 2011; Moulaert et al. 2005; Murray et al. 2010). However, as Moulaert et al. (2005) and Evers et al. (2014) state, caution is expressed on outcomes claiming to 'be good for society' due to their social impact being contextually bounded and subject to interpretation. Innovation is generally understood as a novel solution concerning processes of refunctioning or recombination of existing assets and resources as suggested by BEPA (2011) and Manzini (2015).

Put together, distinct definitions of social innovation are found across disciplines, with varying emphasis. Social innovation is often deemed a quasi-concept with hybrid characteristics adaptable to different situations, flexible for policymakers though also relevant for empirical analysis (e.g. European Commission 2013a, b; BEPA 2014; Bonifacio 2014; TEPSIE 2014). In highlighting the various perspectives, Ardill and Lemes de Oliveira (2018) found 22 different, but complementary, definitions in a comprehensive review. I will now discuss those social innovation definitions I have found most relevant to my research, including the key characteristics and central elements of importance.

There is a split between the two principal meanings of social innovation: one referring to a critical, targeted approach that argues for alternatives to perceived neoliberal agendas, challenging existing power relations and structures, and is presented as 'radical social innovation' (see e.g. Moulaert et al. 2007, 2017; Novy and Hammer 2007; Nicholls et al. 2015; Ayob et al. 2016). The second meaning, presented as 'complementary social innovation', is broader in terms of beneficiaries and aims to meet societal needs more effectively and efficiently than existing solutions (see e.g. Mulgan et al. 2007; Bacon et al. 2008; Murray et al. 2010; BEPA 2011; OECD 2011; Manzini 2014). The complementary perspective is more compatible to existing power relationships and urban governance arrangements as observed by Bonifacio (2014) and Marques et al. (2018). Although authors may not expressly use such terms, they serve to characterise the nature of social innovation being examined.

Of particular interest within radical perspectives is the definition by Moulaert et al. (2005) that was advanced by Moulaert and MacCallum (2019) and permeates across spatial planning, community development, and urban governance perspectives. It stresses three core social innovation dimensions: outcomes in the satisfaction of social needs that are not currently satisfied; processes in changes in social relations, especially regarding bottom-linked territorial governance to increase participation; and empowerment by increasing socio-political capability and access to resources. The implied outcomes of collective empowerment, equality and social justice for equitable development are acknowledged by Evers et al. (2014) and Moulaert et al. (2005) to be subject to social construction.

By way of contrast, complementary definitions tend to fit within existing ways of thinking and work in conjunction with current political systems. They typically aim to increase human capabilities in a societal perspective through the creation of hybrid stakeholder partnerships with end users to 'co-produce' urban public services, spaces and goods more effectively and efficiently than existing solutions to address market failures and state service provision. This human resource view is highlighted by Mulgan (2019) who promotes social innovation as being innovative services that meet social needs by creating cross-sector collaborations in co-production scenarios. This perspective emphasises a product dimension by focussing on replicable initiatives and the social economy for their delivery.

Three central elements of importance can be observed across the different characterisations being proposed by research on social innovation. First, process is routinely expressed as developing new social relations in systems or structures (see e.g. Mumford 2002; Moulaert et al. 2005; Westley and Antadze 2010; Evers et al. 2014; Manzini 2014; TEPSIE 2014). Second, outcomes are presented as concerning the production of social value to meet human needs or addressing societal challenges to sustainable development (see e.g. Mulgan et al. 2007; Bacon et al. 2008; Murray et al. 2010; BEPA 2011; Avelino et al. 2014). Third, the scales of social innovation are discussed as fundamental aspects of their potential transformative impact and engagement of a range of stakeholders (see e.g. Moulaert et al. 2005; Mulgan et al. 2007; Westley and Antadze 2010; de Bruin and Stangl 2013; Avelino et al. 2014; Manzini 2015).

Challenges remain in measuring the longer-term outcomes of social value production due to the conditional nature of social innovations, their context and intended beneficiaries. A task highlighted by assessing the value being produced, whether it is for individuals or to society. Moreover, social value is often presented in such a way that it can be open to interpretation. For example, BEPA (2011) propose that social innovation can result in citizen empowerment, transparency in democracy, social cohesion, and implied spatial justice that will lead to a more productive society.

There are three hierarchical levels of social innovations. At the micro level there are place-based social innovations. They seek to address the social needs of a specific group of citizens and users in a specific location, enhancing capability and promoting wellbeing. They focus on social demand that is characteristically not addressed by market activity or existing organisations and institutions. At the meso level there are social innovations addressing wider societal challenges. They are broader initiatives aimed at an urban to national spatial scale, often concerned with sustainable development practice. They change the development of urban infrastructures, societal sub-systems, communities, organisations, and affect forms of governance. At the macro level there are national to international scale initiatives causing fundamental and sustained changes of existing organisational structures, systems, and relations between institutions and stakeholders. According to Westley and Antadze (2010) and Avelino et al. (2014) these particular social innovations or 'game changers' transform the way in which society thinks and behaves.

3.2 Social Innovation and Sustainable Development

Social innovation connects to the principle of 'sustainable development' through the fulfilment of human needs. Seen by the United Nations as 'meeting the needs of the present without compromising the ability of future generations to meet their own needs' (WCED 1987, p. 43), the 2030 Agenda stresses the importance of innovation to achieving Sustainable Development Goals (United Nations 2015). Given the conceptual focus on satisfying human needs through new social practices and structures, social innovation has the potential to address aspects of the Agenda that emphasise inclusion and equity, markedly for health, education, employment and poverty alleviation. In addressing market or state failures or to incumbent models of service provision, social innovations are seen to challenge existing models of production and consumption through experimenting with potentially transformative, institutional and societal changes. This is significant to attaining the Goals, as they will arguably require marked urban changes. Social innovations therefore provide value as spaces of experimentation with novel ideas and practices, transcending their immediate impact on beneficiaries.

By recognising human development relies on changing social practices, the United Nations has acknowledged that social innovations have significant roles to play in sustainable development. For example, Millard (2018) highlights how their Social Development Network is supporting the use of social innovations in the Asia–Pacific region to help change the attitudes and behaviour of both citizens and service providers to challenges such as public health and access to food resources. To accelerate their diffusion across scale will require cross-sector partnerships and top-down support, especially through state steering. Consequently, this has implications for the multidimensional types of governance arrangements required.

Social innovation being considered as fundamentally good for society can however be a source of contention. Baturina and Bežovan (2015) stress that vulnerable groups can be excluded by initiatives produced by more affluent citizens due to not recognising different group needs in forming public agendas. For instance, the High Line linear park highlights New York's rising inequality and lack of diversity in public spaces. The resident led development sought to transform an abandoned elevated railway to serve a community purpose as a recreational green space within a highly urbanised area (Fig. 3.1). However, the use of public money was controversial. According to Reichl (2016), the initiative received more funding in 2012 than the municipal parks in the rest of the city due to the match funding process. Brenner (2013) suggests it has stimulated real estate development and a process of gentrification in the adjacent Chelsea neighbourhood. Consequently, the nature of social innovations such as the High Line presents challenges for measuring the impact of broader long-term outcomes of social value production and their relationship to socially sustainable development.

Fig. 3.1 The High Line elevated linear park in New York (*Source* Nguyen 2012, CC0 1.0)

The scope of structural change produced by citizen participation is a principal issue in distinguishing between perspectives of social innovation for sustainable development. Radical characterisations target a more inclusive political process on behalf of marginalised groups and deprived communities to direct urban policies towards addressing previously unmet needs. While complementary characterisations do not explicitly specify for changes in socio-political dynamics to accomplish a broader societal aim of meeting human needs.

Approaches within and across social innovation research, whether processes of decision-making, spatial planning, or service provision, highlight the role of active citizen participation. However, to initiate and sustain citizen engagement in sustainable development will require behavioural changes within existing social institutions, structures, and systems. New policies and strategies need to be developed that consider structural variables, such as scale, ownership, and nature of the initiative. Contextual factors such as local leadership and institutional culture will influence the degree of openness and participation. Therefore, social innovation strategies require greater citizen rights in the planning, designing, delivering, and evaluating of urban initiatives in order to support the creation and management of sustainable cities.

3.3 Towards an Understanding on Social Innovation in Social and Urban Change

From the early-2000s new streams of thought on social innovation emerged to confront innovation and social change thinking. This conceptual evolution has directed attention towards changes in cross-sector relationships between government, non-profits, and civil society which create new organisational processes aiming to resolve social needs and problems (e.g. Murray et al. 2010; TEPSIE 2014; Howaldt et al. 2018). There are also perceptible links to co-production approaches which emerged in parallel to the interest on social innovation due to common themes of cross-sector collaboration, idea generation between users and professionals, citizen participation and empowerment, and social change (see e.g. Boyle and Harris 2009; Bovaird and Loeffler 2012; Pestoff 2014; Voorberg et al. 2014).

It is within the above social innovation thought that I position my research. This is because the process of social innovation in the production of urban food spaces concerns the ability of cross-sector stakeholders to 'co-produce' socially valuable urban spaces; the complex institutional settings, differing interests, and conflicts they must overcome, and how this stakeholder agency is situated in evolving social and spatial networks that influence activities. I will highlight the multilevel actions required to develop this nature-based solution in differing urban environments. The socio-ecological approach is widely recognised in its potential yet faces challenges in relation to local contexts of production and deployment as observed by Balian et al. (2016) and Ardill and Lemes de Oliveira (2021). Consequently, the European Commission (2014) has called for more knowledge on common patterns and aggregate learning on the development of social innovation for the environment.

In my research I question the broad conception and scalar perspective of the social innovation pathways offered. Attention is drawn to how social innovation is now viewed as an important component of helping cities transition towards sustainable development. Amid this increasing interest there has been new public funding sources for early stage initiatives and intermediary organisations, the introduction of supportive regulation and legal frameworks, the opening up of access to public assets, and the emergence of global networks to support social innovation. This growth raises questions of the capacity of existing systems and structures in supporting the development of alternative approaches to sustainable change. The Young Foundation (2012) for example observe how social innovations are habitually designed to match to a particular context and produced through a particular arrangement of relationships and collaborations, and therefore developing and replicating these solutions across different geographical settings remains challenging. Christmann (2020) adds that this task is especially evident at the urban scale as spatial policies 'are to a large extent locally situated and bound up with local history' (p. 430).

Social innovation pathways are complex and nuanced. Innovation researchers contend that spatial scales and contextual dynamics perform a significant function. This influence is particularly evident in government policy and strategies at national to urban scales, though also following periods of social upheaval or crisis, through the emergence of social movements, and instability in systems. According to The Young Foundation (2012) and Moulaert et al. (2017), this process creates openings for innovation in shaping societal modes of production and consumption to confront current systems. Yet even with this recognition, interest centres on the scaling of social innovation and replication and transfer of identikit strategies from one region to another.

I therefore agree with Cajaiba-Santana (2014) in that in order to move the field forward a process framework is required to understand the complexities of actions taken in environmental and time dependent settings typical for social innovations. Currently lacking is an increased awareness of the development of social innovation in urban situations as a complex and multi-layered process. In this sense, a process research approach is called for where a discreet set of change events in a specific location are analysed based on a historical narrative that will reconstruct series of chronological sequences of events. Such an approach is suggested by Van de Ven and Poole (2004) to enable the identification of processes 'through which change unfolds over time' (p. 296).

References

Ardill N, Lemes de Oliveira F (2018) Social innovation in urban spaces. Int J Urban Sustain Dev 10(3):207–221. https://doi.org/10.1080/19463138.2018.1526177

Ardill N, Lemes de Oliveira F (2021) Emerging places of social innovation (POSI): a conceptual framework for social innovation in cities. Trans Assoc Eur Sch Plan 5(1):55–70. https://doi.org/10.24306/TrAESOP.2021.01.005

Avelino F, Wittmayer J, Haxeltine A et al (2014) Game changers and transformative social innovation. TRANSIT, Brussels

Ayob N, Teasdale S, Fagan K (2016) How social innovation "came to be": tracing the evolution of a contested concept. J Soc Policy 45(4):1–19

Bacon N, Faizullah N, Mulgan G, Woodcraft S (2008) Transformers. How local areas innovate to address changing social needs. NESTA, London

Balian E, Berhault A, Eggermont H et al (2016) Social innovation and nature-based solutions. EKLIPSE, Brussels

Baturina D, Bežovan G (2015) (Social) Innovation impact—review of research. Third Sector Impact, Brussels

Benneworth P, Cunha J (2015) Universities contributions to social innovation: reflections in theory and practice. Eur J Innov Manag 18(4):508–527

BEPA (2011) Empowering people, driving change social innovation in the European Union. European Commission, Brussels

BEPA (2014) Social innovation a decade of changes. European Commission, Brussels

Bonifacio M (2014) Social innovation: a novel policy stream or a policy compromise? An EU perspective. Eur Rev 22(1):145–169

Bovaird T, Loeffler E (2012) From engagement to co-production: the contribution of users and communities to outcomes and public value. Voluntas 23(4):1119–1138

Boyle D, Harris M (2009) The challenge of co-production. NESTA, London

Brenner N (2013) Open city or the right to the city? TOPOS Int Rev Landsc Archit Urban Des 85(1):42–45

Cajaiba-Santana G (2014) Social innovation: moving the field forward. A conceptual framework. Technol Forecast Soc Change 82(1):42–51
Christmann GB (2020) Introduction: struggling with innovations. Social innovations and conflicts in urban development and planning. Eur Plan Stud 28(3):423–433
de Bruin A, Stangl LM (2013) The social innovation continuum: towards addressing definitional ambiguity. In: 4th EMES international research conference on social enterprise. EMES, Liege
European Commission (2013a) Guide to social innovation. European Commission, Brussels
European Commission (2013b) Social innovation research in the European Union. European Commission, Brussels
European Commission (2014) Social innovation and the environment. European Commission, Brussels
Evers A, Ewert B, Brandsen T (eds) (2014) Social Innovations for social cohesion. WILCO, Liege
Howaldt J, Kaletka C, Schröder A, Zirngiebl M (eds) (2018) Atlas of social innovation. TU Dortmund University, Dortmund
Manzini E (2014) Making things happen: social innovation and design. MIT Des Issues 30(1):1–12
Manzini E (2015) Design, when everybody designs: an introduction to design for social innovation. MIT Press, Cambridge, MA
Marques P, Morgan K, Richardson R (2018) Social innovation in question: the theoretical and practical implications of a contested concept. Environ Plan C 36(3):496–512
Millard J (2018) How social innovation underpins sustainable development. In: Howaldt J, Kaletka C, Schröder A, Zirngiebl M (eds) Atlas of social innovation. TU Dortmund University, Dortmund, pp 40–43
Moulaert F, MacCallum D (2019) Advanced introduction to social innovation. Edward Elgar Publishing, Cheltenham
Moulaert F, MacCallum D, Mehmood A, Hamdouch A (eds) (2013) The international handbook on social innovation. Edward Elgar, Cheltenham
Moulaert F, Martinelli F, González S, Swyngedouw E (2007) Social innovation and governance in European cities. Eur Urban Reg Stud 14(3):195–209
Moulaert F, Martinelli F, Swyngedouw E, Gonzalez S (2005) Towards alternative model(s) of local innovation. Urban Stud 42(11):1969–1990
Moulaert F, Mehmood A, MacCallum D, Leubolt B (eds) (2017) Social innovation as a trigger for transformations. European Commission, Brussels
Mulgan G (2019) Social innovation: how societies find the power to change. Policy Press, Bristol
Mulgan G, Tucker S, Ali R, Sanders B (2007) Social innovation: what it is, why it matters, how it can be accelerated. Skoll Centre for Social Entrepreneurship, Oxford
Mumford MD (2002) Social innovation: ten cases from Benjamin Franklin. Creat Res J 14(2):253–266
Murray R, Caulier-Grice J, Mulgan G (2010) The open book of social innovation. NESTA, London
Nguyen D (2012) High line park. https://www.flickr.com/photos/zokuga/7860320800/. Accessed 25 Nov 2020
Nicholls A, Simon J, Gabriel M (eds) (2015) New frontiers in social innovation research. Palgrave Macmillan, Basingstoke
Novy A, Hammer E (2007) Radical innovation in the era of liberal governance. Eur Urban Reg Stud 14(3):210–222
OECD (2011) Fostering innovation to address social challenges. OECD, Paris
Pestoff V (2014) Collective action and the sustainability of co-production. Public Manag Rev 16(3):383–401
Phills J, Deiglmeier K, Miller D (2008) Rediscovering social innovation. SSIR 6(4):34–43
Pol E, Ville SP (2009) Social innovation: buzz word or enduring term? J Socio Econ 38(6):878–885
Reichl AJ (2016) The high line and the ideal of democratic public space. Urban Geogr 37(6):904–925
Rüede D, Lurtz K (2012) Mapping the various meanings of social innovation: towards a differentiated understanding of an emerging concept. EBS Business School, Oestrich-Winkel

TEPSIE (2014) Social innovation theory and research: a summary of the findings from TEPSIE. European Commission, DG Research, Brussels

The Young Foundation (2012) Social innovation overview. European Commission, Brussels

United Nations (2015) Transforming our world: the 2030 agenda for sustainable development. United Nations Environment Programme, New York

Van de Ven A, Poole M (eds) (2004) Handbook of organizational change and innovation. Oxford University Press, New York

Voorberg W, Bekkers V, Tummers L (2014) A systematic review of co-creation and co-production: embarking on the social innovation journey. Public Manag Rev 17(9):1333–1357

WCED (1987) Our common future. United Nations, Geneva

Westley F, Antadze N (2010) Making a difference: Strategies for scaling social innovation for greater impact. Innov J Public Sect Innov J 15(2):1–19

Chapter 4
Social Innovation Process in the Production of Urban Spaces

Abstract This chapter presents a conceptual framework to investigate social innovation dynamics in the production of spaces for food growing in cities. The framework combines social innovation research from process theories and discussion on urban spaces to provide a basis for interpreting how processes of change have occurred. The chapter begins by analysing and breaking down existing process models of social innovation into their core components and proposes integration of the main dynamics: the progress of the innovation, changes in social relations, and collaborative agency. It then constructs a conceptual framework comprising of four key processes: (1) Identification of human needs or societal challenges to sustainable development; (2) Development of social relations in systems or structures; (3) Provision of opportunity for social empowerment; (4) Reflection of socio-spatial development practice. The proposed framework draws attention to the importance of partnership working and intermediation activities in order to improve access to public spaces, whilst contributing to social justice and healthy orientated urban environments.

Keywords Social innovation processes · Co-production of space · Socio-spatial dynamics · Social action · Social innovation · Urban development

4.1 Process Models of Social Innovation

In this section I present a framework to investigate the process of social innovation in the production of urban food growing spaces after reviewing three existing models of social innovation. Mulgan (2006) proposed a framework for the process of social innovation that was advanced by Murray et al. (2010) and other collaborators from

---

This chapter draws from previously published: Ardill & Lemes de Oliveira (2019) Emerging Places of Social Innovation (POSI). The co-production of space between multilevel stakeholders: the case of productive urban green infrastructure. In AESOP (Ed.) *Planning for Transition* (pp. 2303–2319). Venice: Association of European Schools of Planning;

Ardill & Lemes de Oliveira (2021) Emerging Places of Social Innovation (POSI): a Conceptual Framework for Social innovation in Cities, *Transactions of the Association of European Schools of Planning*, 5:1, 55–70, https://doi.org/10.24306/TrAESOP.2021.01.005.

N. Ardill, *Growing Food in Cities*, Cities and Nature,
https://doi.org/10.1007/978-3-030-98475-5_4

the Young Foundation and NESTA UK (e.g. Mulgan et al. 2007; Bacon et al. 2008; SIX 2010; Caulier-Grice et al. 2012), and identified six stages:

(1) Prompts, inspirations, and diagnoses (which involves identifying and defining a need to be met)
(2) Proposals and ideas (the stage of idea generation and designing ways to deal with the identified need)
(3) Prototyping and pilots (where ideas get tested in practice through pilot projects with feedback from users and experts)
(4) Sustaining (when the idea becomes everyday practice)
(5) Scaling and diffusion (which involves developing a range of strategies for growing and spreading an innovation to a larger group or to other communities)
(6) Systemic change (so that it works on a broader scale by introducing entire systems)

Sustainable systemic changes in redesigning society through changes in the relationships that exist between institutions and stakeholders is positioned by policy advisors as being the principal focus of social innovation as expanded by Murray et al. (2010); SIX (2010); BEPA (2011); and Baturina and Bežovan (2015). Commenting further, Caulier-Grice et al. (2012) highlight that the innovation process proposed is repetitive rather than linear, and that the model should be considered as multiple spirals rather than straight lines (Fig. 4.1). Therefore, it should not be assumed that initiatives will go through all six stages as many will jump between or skip entire stages altogether. In fact Caulier-Grice et al. (2012) state that some cases of social innovation 'remain small in scale and locally based, rather than attempting growth and scale, and very few social innovations effect or reach the stage of systemic change' (p. 34). It follows, that social innovation is understood broadly as the production

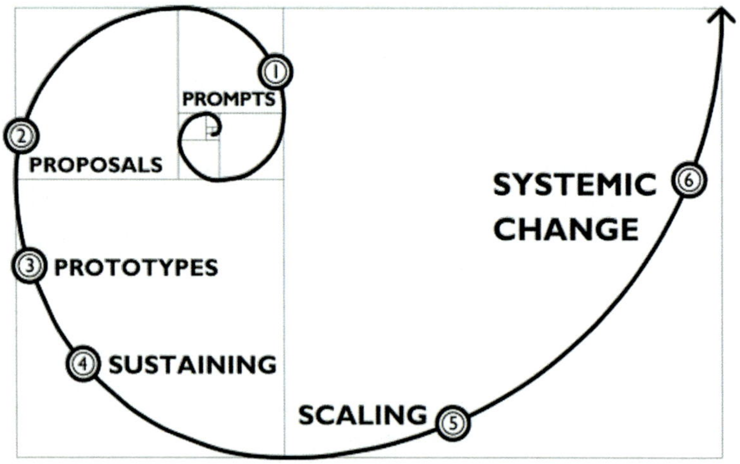

Fig. 4.1 The six stages of social innovation as formulated by the Young Foundation and NESTA UK (Adapted from Murray et al. 2010)

of societal value in meeting social needs and creating new social relationships or collaborations to enhance society's capacity to act (see, e.g. Mulgan 2006; Mulgan et al. 2007; Murray et al. 2010).

Moulaert et al. (2005, 2010) has put forward another framework, the Alternative Model of Local Innovation by to counter the social exclusion dynamics experienced at various spatial scales. The model conceptualises social innovation dynamics which occur in interaction with each other over time, beginning with the deprivation of human needs across four areas: economic and/or material basic needs, such as food, clothing, shelter and employment; social needs of health and education; cultural needs of self-expression, identity and recognition; and political needs of equal opportunity and active citizenship. The deprivation of needs causes a reaction and the mobilisation of resources. These resources are recognised as human, social and institutional, organisational, and financial. Mediation is required between stakeholders (e.g. civil society and state) in order to develop social initiatives which satisfy those human needs not currently being satisfied (Fig. 4.2). This activity fosters processes of social changes in existing social and power relations towards inclusive and democratic urban governance systems to 'increase the level of participation of all but especially deprived groups in society' (Moulaert et al. 2005, p. 1976). As proposed by Moulaert et al. (2005, 2010), this sequence enables previously excluded social groups to be empowered through increasing their socio-political capability and access to the resources needed to improve rights to satisfaction of human needs and participation. Thus, social innovation is understood from a radical perspective as the

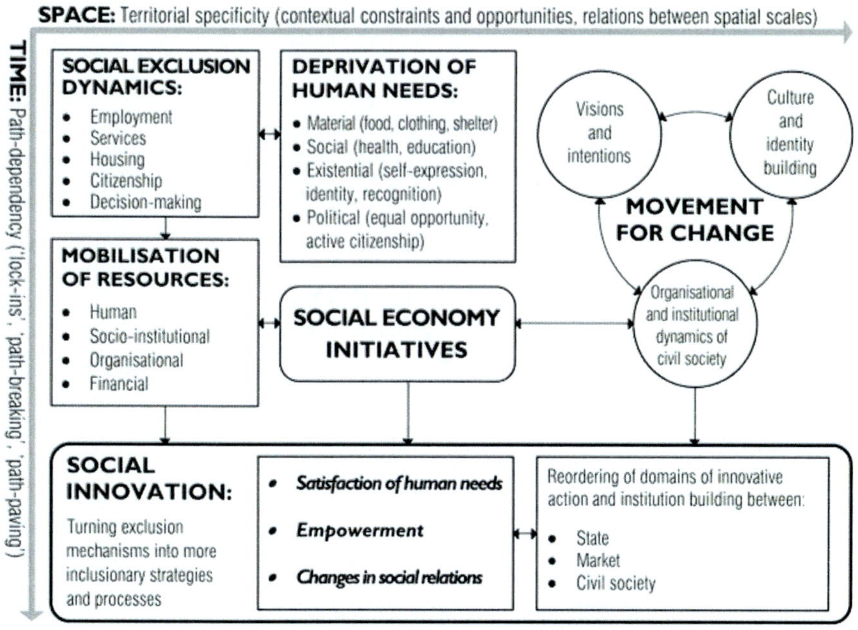

Fig. 4.2 The Alternative Model of Local Innovation (Adapted from Moulaert et al. 2005, 2010)

social changes that achieve conditions of empowerment, and favour bottom-linked governance initiatives and inclusive urban development, and 'explicitly refers to an ethical position of social justice' (Moulaert et al. 2005, p. 1978).

A further model is drawn by Ayob, Teasdale, and Fagan (2016). Examining how the concept has developed over time, they argue that the social innovation process has 'five plausible routes through some or all of this process, all of which can be conceived of as social innovation' (Ayob et al. 2016, p. 648). The five routes are:

(1) New forms of social relations lead to innovation
(2) Innovation leads to a restructuring of social and or power relations
(3) Innovation leads to utilitarian social value
(4) New forms of social relations lead to innovation which results in the restructuring of power relations (and thus societal impact)
(5) New forms of social relations lead to innovation, which creates utilitarian social value (and thus societal impact)

In doing so, the authors distinguish between two social innovation approaches and outcomes in social change. The first, seen as utilitarian, emphasises changes in 'aggregate individual utility'. Or to use other words, the sum of useful gains for people positively impacted, less the total experienced by those negatively impacted by the innovation. The second, considered more radical, 'sees social (and political) change occurring as a consequence of innovations in social relations' (Ayob et al. 2016, p. 648).

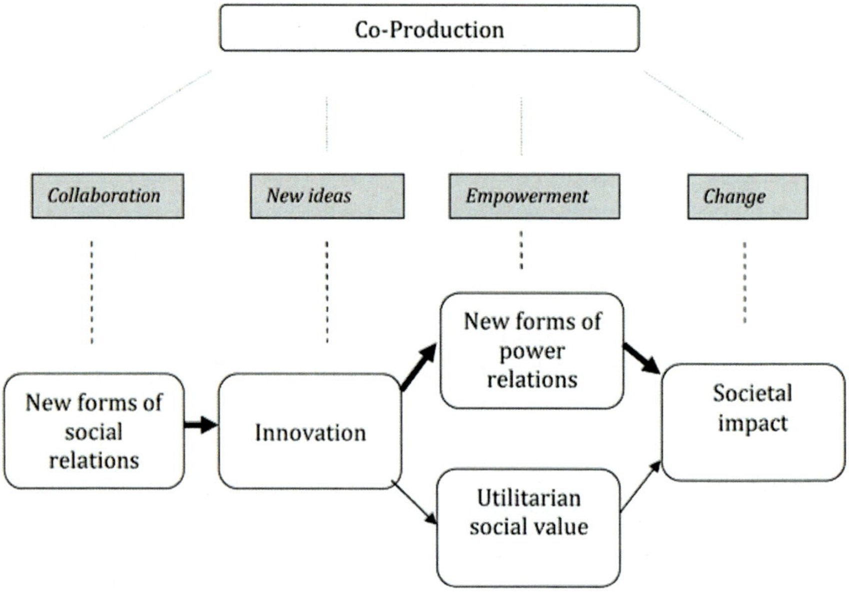

Fig. 4.3 Social innovation pathways and drawing the correlation to co-production (*Source* Ayob et al. 2016)
Note 'Strong social innovation' pathway indicated by bold arrows

The authors proceed to draw similarities between co-production and the five social innovation pathways outlined (see Fig. 4.3). The radical approach, termed 'strong social innovation', is suggested in this model to be strongly linked to co-production due to their shared emphasis on shifting power influences and dynamics between citizens and public institutions as a key component of this approach, notably through the engagement and empowerment of previously disadvantaged groups (Ayob et al. 2016).

In summary, the outlined models attempt to address different questions and develop their own viewpoints on social innovation process. Murray et al. (2010), building on the work of Mulgan (2006) and fellow collaborators, are interested by how innovations in the social field progress, and identified six stages 'that take ideas from inception to impact' (Murray et al. 2010, p. 12). Whereas, Moulaert et al. (2005, 2010) examined what structural changes in social relations are happening, and make connections between urban governance, empowerment, and social justice. Finally, Ayob et al. (2016) explored how social innovation has evolved, and in so doing linked pathways to co-production and shared actions in developing collaborative forms of social relations, leading to changes and societal impact.

Table 4.1 Relationship between different stages of the social innovation process models

<table>
<tr><th>Murray et al. (2010)
How innovation progresses?</th><th>Moulaert et al. (2005, 2010)
What innovation in social relations is taking place?</th><th>Ayob et al. (2016)
How do collaborative actions cause change?</th><th>Ardill and Lemes de Oliveira (2019)
What is the stage of the particular process?</th></tr>
<tr><td>Prompts, inspirations and diagnoses</td><td rowspan="2">Deprivation of human needs</td><td>New forms of social relations</td><td rowspan="2">*Identification of human needs or societal challenges to sustainable development*</td></tr>
<tr><td rowspan="3">Proposals and ideas</td><td rowspan="3">Innovation</td></tr>
<tr><td>Mobilisation of resources</td><td rowspan="2">*Development of social relations in systems or structures*</td></tr>
<tr><td>Changes in social relations (and political relations)</td></tr>
<tr><td>Prototyping and pilots</td><td rowspan="2">Empowerment</td><td>Utilitarian social value</td><td>*Provision of opportunity for social empowerment*</td></tr>
<tr><td>Sustaining</td><td>New forms of power relations</td><td rowspan="3">*Reflection of socio-spatial development practice*</td></tr>
<tr><td>Scaling and diffusion</td><td rowspan="2">Satisfaction of human needs and participation</td><td rowspan="2">Societal impact</td></tr>
<tr><td>Systemic change</td></tr>
</table>

(*Source* Author)

In my study, I integrate the three innovation models into a social innovation framework that encompasses the following three components: the progress of the innovation, changes in social relations, and collaborative dynamics. I do so by defining a model of the process of social innovation in the production of urban spaces involving four cyclical stages. Table 4.1 indicates how the stages of this model relate to the elements in the Murray et al. (2010), Moulaert et al. (2005, 2010), and Ayob et al. (2016) models.

4.2 A Social Innovation Framework for Understanding Key Processes in the Production of Urban Spaces

The traditional linear process of innovation assumes that innovation always starts with research, is then followed by development, and ends with production and diffusion (see, e.g. Godin 2006; Balconi et al. 2010). This model has been much criticised and fails to recognise that 'knowledge does not flow smoothly among different stages of the innovative process and among different organizations and institutions. Nor does it flow freely among geographical areas' (Balconi et al. 2010, p. 7).

Following Murray et al. (2010), I propose a framework for social innovation process that is based on cyclic innovation processes, which allows for overlap, interaction and nonlinearity, as 'change needs to be understood through the iterative action of the processes and dynamics' (Van de Ven and Poole 2004, p. 317). While the individual stages are not always necessarily linear or sequential, I identify four key stages in providing an analytical framework by which to think through all the activities taking place, the various agents involved, and the patterns which occur in the context of such innovation journeys (Ardill and Lemes de Oliveira 2021). The framework I am proposing is visualised in Fig. 4.4 as a circular process; the implication being that social and spatial change is a constant activity. The key stages combine insights from social innovation research and are described in the following paragraphs.

The first stage: 'Identification of human needs or societal challenges to sustainable development' involves prompts that highlight the need for innovation to address human needs or societal challenges as demonstrated by Murray et al. (2010) and SIX (2010). Human needs may include the basic or specific needs of individuals and groups. For example, Maslow (1954) characterises the basic needs of physiological needs (e.g. food, clothing, and shelter), and safety needs as those relating to health and wellbeing, employment, and security. Whereas, societal challenges viewed from a sustainable development perspective are directed towards society as a whole and are recognised as major concerns that are shared by all citizens; especially uneven development, health, and climate action (BEPA 2011; Ardill and Lemes de Oliveira 2018). These grand challenges are highlighted by people and space interactions. Long-standing and emerging urban problems are brought into focus here by

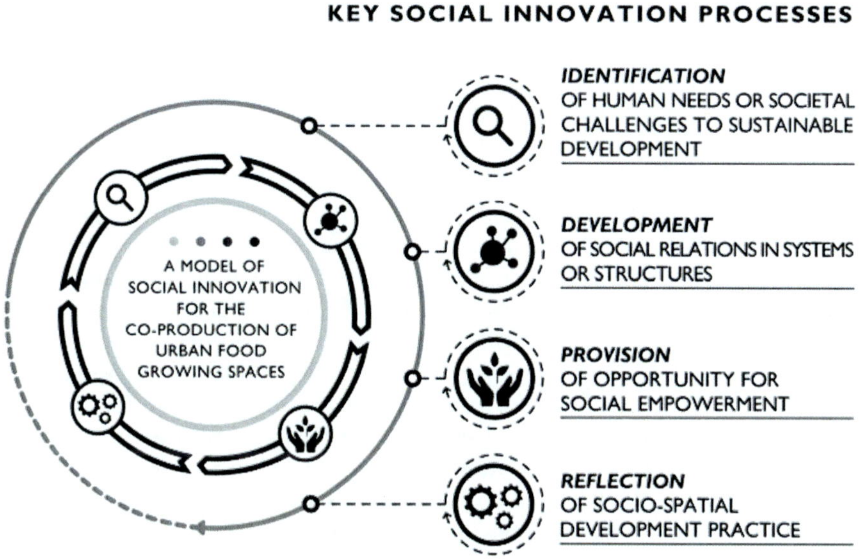

Fig. 4.4 Proposed social innovation framework for understanding key processes in the production of urban spaces (*Source* Author)

an experience or event, or through the research and interpretation of a social agenda by initiators of the innovation (Murray et al. 2010).

The identification process involves diagnosing unmet human needs or societal challenges by understanding the context dynamics (e.g. social, economic, physical etc.) which affect a given situation in order to assess and frame opportunities and constraints for innovation (SIX 2010). From the identification of needs an idea for a solution is generated. Evidence gathered is combined as findings, and made into a persuasive argument to stakeholders affected by the innovation that the solution proposed can be effective, and a defined brief for development with strategic objectives and directions is set out (Torresa 2017).

The second stage is the 'Development of social relations in systems or structures'. A multitude of stakeholders will typically be engaged in this stage. For example, these stakeholders could be the ones that have identified the needs or challenges and other stakeholders that are interested in, or might directly benefit from, addressing the specific socio-spatial agenda. Generating cross-sector support and coordination is valuable here to mobilise the resources needed to work on the social innovation solution and the co-production approach presents a way of collaborative working (see, e.g. Boyle and Harris 2009; Voorberg et al. 2014; Ayob et al. 2016). The setup of a coalition and supportive structures further develop the innovative solution, and the creation of a protected space for experiment are significant features of this process (Rip 2012). This approach is aided by innovation intermediaries, such as agents and organisations which create opportunities and innovation spaces that can be social, economic, and physical. Stewart and Hyysalo (2008) propose that intermediation will require facilitation, configuring, and brokering activities which create social relationships to support the planned innovation.

The contribution of civil society through social entrepreneurship and enterprise aligned with state steering to coordinate processes of social innovation creates the conditions required for hybrid stakeholder partnerships to emerge as understood by Mulgan et al. (2007), Baker and Mehmood (2015), and Howaldt et al. (2018). New coalitions are comprised of public, private, and social participants in the organisation of urban development. Mumford (2002) and Moulaert et al. (2005) suggest that these coalitions contribute to the rearrangement and restructuring of existing social relationships. Furthermore, the use of embedded resources and assets is a way of engaging a range of stakeholders in the co-design and development of solutions, whilst design-based approaches fusing design-thinking can progress and shape the idea (see, e.g. Brown and Wyatt 2010; Manzini 2015). To help ensure that needs are met, collaboration amongst the stakeholders contributing to the development of the social innovation solution is considered significant by Voorberg et al. (2014).

The third stage: 'Provision of opportunity for social empowerment' is where a spatial initiative is implemented, and seeks to create openings to enhance society's capacity to act in a changing environment (Murray et al. 2010; BEPA 2011; TEPSIE 2014). This involves generating social value, both to disadvantaged groups and society as a whole, and through increasing participation in urban governance structures to increase access to public resources (see, e.g. Gerometta et al. 2005; Moulaert et al. 2005; Ayob et al. 2016; Brandsen et al. 2016).

In this delivery stage, opportunities for community development are enabled through inclusive practices and social engagement which encourage active citizenship to help meet needs. Social learning activities, where people can learn from each other collectively rather than through isolated individual activities, will increase community capacity through the development of new skills to construct more resilient communities (Pol and Ville 2009; Reed et al. 2010; Manzini 2015). This building of capacities is important to contribute to sustainable placemaking and the promotion of sustainable development (Mehmood and Parra 2013; Baker and Mehmood 2015). Overall, improving access to urban resources helps to build capacities, whilst changes in group-decision-making and power relations will create new socio-political capabilities, and enhance peoples control over their own lives to support social justice (Moulaert et al. 2005, 2010; MacCallum et al. 2009).

The fourth stage connecting the process cycle is 'Reflection of socio-spatial development practice'. This stage involves considering measures of the success of the initiative, as well as the processes of selecting, developing, and prescribing a model of standardisation as suggested by Murray et al. (2010); SIX (2010); and Bund et al. (2015). The activities of demonstrating, refining, and testing ideas to obtain feedback from users and specialists in order to evolve solutions and maximise impact are important to learning. Through repetition conflicts can be resolved and coalitions of stakeholders gather strength. Hillgren et al. (2011) and Bjögvinsson et al. (2012) argue that this activity supports the creation of structures that enable participation in spatial development and helps to embed local stakeholder relations, networks, and resources.

The adaption of the idea and sustaining the initiative through use of evidence and identifying further resources is necessary if the innovation is to be carried forward. While the spreading and sharing of the solution through diffusion and emulating occurs, it follows that the provision of support and know-how from one organisation or place to another is significant to open knowledge advancement (see, e.g. Caulier-Grice et al. 2012; Chesbrough et al. 2014; TEPSIE 2014). As Windrum et al. (2016) and Torresa (2017) state, this process is necessary to move innovation from a community level to a widespread solution. It is important therefore to identify how an initiative can be translated and implemented in other contexts so that it can provide solutions in new situations and places.

References

Ardill N, Lemes de Oliveira F (2018) Social innovation in urban spaces. Int J Urban Sustain Dev 10(3):207–221. https://doi.org/10.1080/19463138.2018.1526177

Ardill N, Lemes de Oliveira F (2019) Emerging places of social innovation (POSI). The co-production of space between multilevel stakeholders. In: "Planning for transition", annual congress for the association of European schools of planning (AESOP), Aesop 2019 conference, Venice, 9–13 July. AESOP, Venice, pp 2303–2319

Ardill N, Lemes de Oliveira F (2021) Emerging places of social innovation (POSI): a conceptual framework for social innovation in cities. Trans Assoc Eur Sch Plan 5(1):55–70. https://doi.org/10.24306/TrAESOP.2021.01.005

Ayob N, Teasdale S, Fagan K (2016) How social innovation "came to be": tracing the evolution of a contested concept. J Soc Policy 45(4):1–19

Bacon N, Faizullah N, Mulgan G, Woodcraft S (2008) Transformers. How local areas innovate to address changing social needs. NESTA, London

Baker S, Mehmood A (2015) Social innovation and the governance of sustainable places. Local Environ Int J Justice Sustain 20(3):321–334

Balconi M, Brusoni S, Orsenigo L (2010) In defence of the linear model: an essay. Res Policy 39(1):1–13

Baturina D, Bežovan G (2015) (Social) innovation impact—review of research. Third Sector Impact, Brussels

BEPA (2011) Empowering people, driving change Social Innovation in the European Union. European Commission, Brussels

Bjögvinsson E, Ehn P, Hillgren P-A (2012) Design things and design thinking: contemporary participatory design challenges. Des Issues 28(3):101–116

Boyle D, Harris M (2009) The challenge of co-production. NESTA, London

Brandsen T, Cattacin S, Evers A, Zimmer A (eds) (2016) Social innovations in the Urban context. Springer, London

Brown T, Wyatt J (2010) Design thinking for social innovation. SSIR 8(1):31–35

Bund E, Gerhard U, Hoelscher M, Mildenberger G (2015) A methodological framework for measuring social innovation. Hist Soc Res 40(3):48–78

Caulier-Grice J, Davies A, Patrick R, Norman W (2012) Defining social innovation. European Commission, DG Research, Brussels

Chesbrough H, Vanhaverbeke W, West J (2014) New frontiers in open innovation. Oxford University Press, Abingdon

de Arruda Torresa PM (2017) Design for socio-technical innovation: a proposed model to design the change. Des J 20(sup1):S3035–S3046

Gerometta J, Haussermann H, Longo G (2005) Social innovation and civil society in Urban governance: strategies for an inclusive city. Urban Stud 42(11):2007–2021
Godin B (2006) The linear model of innovation: the historical construction of an analytical framework. Sci Technol Hum Values 31(6):639–667
Hillgren PA, Seravalli A, Emilson A (2011) Prototyping and infrastructuring in design for social innovation. CoDesign 7(3–4):169–183
Howaldt J, Kaletka C, Schröder A, Zirngiebl M (eds) (2018) Atlas of social innovation. TU Dortmund University, Dortmund
MacCallum D, Moulaert F, Hillier J, Haddock SV (eds) (2009) Social innovation and territorial development. Ashgate, Farnham
Manzini E (2015) Design, when everybody designs: an introduction to design for social innovation. MIT Press, Cambridge, Massachusetts
Maslow A (1954) Motivation and personality. Harper, New York
Mehmood A, Parra C (2013) Social innovation in an unsustainable world. In: Moulaert F, MacCallum D, Mehmood A, Hamdouch A (eds) The international handbook on social innovation: collective action, social learning and transdisciplinary research. Edward Elgar, Cheltenham, pp 53–66
Moulaert F, Martinelli F, Swyngedouw E, Gonzalez S (2005) Towards alternative model(s) of local innovation. Urban Stud 42(11):1969–1990
Moulaert F, Martinelli F, Swyngedouw E, Gonzalez S (eds) (2010) Can neighbourhoods save the city? Community development and social innovation. Routledge, Abingdon
Mulgan G (2006) The process of social innovation. Innov Technol Govern, Glob 1(2):145–162
Mulgan G, Tucker S, Ali R, Sanders B (2007) Social Innovation: what it is, why it matters, how it can be accelerated. Skoll Centre for social entrepreneurship, Oxford
Mumford MD (2002) Social innovation: ten cases from Benjamin Franklin. Creat Res J 14(2):253–266
Murray R, Caulier-Grice J, Mulgan G (2010) The open book of social innovation. NESTA, London
Pol E, Ville SP (2009) Social innovation: buzz word or enduring term? J Socio Econ 38(6):878–885
Reed M, Evely A, Cundill G, et al (2010) What is social learning? Ecol Soc 15(4):r1
Rip A (2012) The context of innovation journeys. Creat Innov Manag 21(2):158–170
SIX (2010) Study on social innovation. The Young Foundation, London
Stewart J, Hyysalo S (2008) Intermediaries, users and social learning in technological innovation. Int J Innov Manag 12(3):295–325
TEPSIE (2014) Social innovation theory and research: a summary of the findings from TEPSIE. European Commission, DG Research, Brussels
Van de Ven A, Poole M (eds) (2004) Handbook of organizational change and innovation. Oxford University Press, New York
Voorberg W, Bekkers V, Tummers L (2014) A systematic review of co-creation and co-production: embarking on the social innovation journey. Public Manag Rev 17(9):1333–1357
Windrum P, Schartinger D, Rubalcaba L et al (2016) The co-creation of multi-agent social innovations. Eur J Innov Manag 19(2):150–166

Part III
Case Studies of Social Innovation in Urban Food Spaces

Chapter 5
Social Innovation in Urban Food Spaces: An Overview of European Cities

Abstract This chapter presents an overview of European activity around social innovation in urban food spaces by drawing on case studies from 12 European cities from all four European geographic regions to enable a comparative view of practice. These studies connect to the social and spatial dynamics being investigated, expressly, urban governance, civic participation, and collaborative approaches to addressing social needs. This survey focuses on experiences in European cities, where social and cultural conditions are more similar to the UK conditions than outside of Europe, and therefore are more comparable to a UK context (analysed in Chaps. 6–11). In selecting cases, emphasis is placed on ensuring variation by choosing urban food spaces in different national contexts with different traditions of spatial practice and civic participation. Cases are examined across cities with differences in history, demographics, density, political governance, and social challenges. The findings indicate that in general the mobilisation of local inhabitants is significant to the process of social innovation in urban development, whilst an active citizenship can support a reproduction of innovation to increase participation in urban decision-making processes.

Keywords Community gardens · Social innovation · Co-production of space · Civic participation · Urban Governance · Urban food growing · Green space · Public space · European cities

In this chapter, I present an overview of case studies of social innovation in the production of community food gardens and orchards in European cities. While these gardening initiatives share in common a collective spatial practice, the initiatives have different aims and social agendas, their organisational structure varies, and the urban spaces produced have a different character. There are gardening initiatives where the main purpose is improving public access to green space, initiatives where educational activity is the focal point, and initiatives where it is more about encouraging social integration and interaction by giving local residents an opportunity to meet their neighbours. Some of these cultivation spaces are located on former industrial waste-lands in densely populated urban districts, some are developed within public parks or squares, and others are integrated into urban planning and development. The cases examined range from well-established projects with institutional support of local

N. Ardill, *Growing Food in Cities*, Cities and Nature,
https://doi.org/10.1007/978-3-030-98475-5_5

Fig. 5.1 Location of 12 case studies in urban food spaces in European cities and regions (*Source* Author)

authorities and non-governmental organisations to recently emerging and loosely organised initiatives of a temporary nature. Together, the 12 case studies serve to illustrate the many opportunities and challenges to social innovation in cultivating urban spaces in European cities and regions (see Fig. 5.1 for locations).

5.1 Urban Food Growing: Case Studies of Social Innovation in European Cities

My survey emphasises different organisational activities and cooperation among city stakeholders occurring at different spatial scales. The stakeholder agency advancing social innovation in collaborative spatial production can be categorised into three types: (1) Neighbourhood level, bottom-up approaches to community based innovation in meeting local residents' needs; (2) Non-governmental organisation, intermediate/ hybrid approaches to practitioner based innovation in developing food growing programmes with socio-ecological objectives; and (3) Citywide, or institutional, top-down approaches to policy based innovation in addressing societal challenges around sustainable development.

Overview of Urban Food Growing Case Studies from Western Europe Cites

Prinzessinnengarten, Berlin

In Berlin, Germany, the not-for-profit organisation Common Grounds coordinates a community garden ('*Prinzessinnengarten*'), which is addressing issues of food growing, public space and social justice in the Kreuzberg district, where municipal land and property is threatened by privatisation. The rapid increase in Berlin land prices stimulated by the economic recovery after the 2008 financial crisis caused urban gardening projects on former industrial wastelands to face displacement. Contextually, there are more than 100 such gardens in the city (Hartmann 2019), with many developed after the fall of the Berlin Wall in 1989; exposing large areas of urban land in addition to the integration of undeveloped land from East Berlin. This emerging urban space was soon occupied by citizens and community groups both legally and illegally in claiming a Lefebvrian 'right to the city', re-producing unused urban spaces as appropriated social spaces (see, e.g. Lefebvre 1991).

In 2009, the *Prinzessinnengarten* was launched as a pilot initiative by two social entrepreneurs alongside a volunteer group of local residents and urban activists on wasteland at Moritzplatz rented from the City of Berlin (Fig. 5.2). The site is 0.6 ha in area and comprises a mobile garden in containers to allow for possible relocation to another site. The primary aim is to make the garden a place of informal learning through its social, educational, and economic activities while engaging people in wider debates on the alternative use of urban land. In particular, the promotion of socially innovative solutions to urban welfare, ecological cultivation, and sustainable development. This social enterprise is self-financed through selling produce to the

Fig. 5.2 Berlin-Kreuzberg urban wasteland at Moritzplatz before and after the development of *Prinzessinnengarten* (*Source* Prinzessinnengarten 2019, CC BY-SA 3.0)

public, an on-site garden café, and by developing over 40 additional gardens for kindergartens, schools, and other local institutions. It is estimated that around 60,000 Berliners visit the garden each year, with up to 1,000 volunteers supporting food growing activities throughout the season (Prinzessinnengarten 2019).

The *Prinzessinnengarten's* development as a significant food growing area and transformation of former wasteland in the centre of the city shows that collaboration and collective learning among a diverse and active urban citizenship is achievable. However, the initiative also reveals an uncertainty of grassroots social innovations around urban food growing as the site has faced closure on several occasions. For instance, in 2012, upon expiry of a three-year lease the public land was due to be sold to a private investor, only prevented due to a petition signed by 30,000 residents, and in 2019 its future was uncertain after a second land lease expired. At the time, the revised ambition for the initiative was to secure municipal funding and a 99-year lease to protect against commercial development (Prinzessinnengarten 2019). As a result, the *Prinzessinnengarten* demonstrates a common challenge to the long-term sustainability of many such urban community gardens in only having a 'temporary use' contract before vacant land is sold and redeveloped.

De Bikkershof, Utrecht

Utrecht, like many cities in the Netherlands, is challenged by a high urban density and relatively few public areas. To address this problem, the city's planning department has developed a novel strategy for its public spaces that is based on collaboration with citizens. The strategic aim is to promote citizens' participation in the development of the city and its maintenance to facilitate opportunities that will improve the quality of life for its citizens. In practice, this means that Utrecht has developed a strong reputation in the Netherlands for citizen participation in urban community gardens and in its green and public spaces (Utrecht Natuurlijk 2016).

'De Bikkershof' ('The Biker Court') in the Wittevrouwen district to the northeast of the city was one of the first community food gardens and self-management projects established in Utrecht, becoming a template for future projects. Launched in 1987, this citizen-managed garden on municipal owned land is 0.3 ha in area, based on permaculture principles and comprises of orchards, vegetable gardens, animals and a playground for children (Bewonersvereniging Bikkershof 2016). The initiative has won awards for its attention to education and guided tours are given to interested groups from across Europe in addition to lectures on ecological living and urban food growing to disseminate knowledge.

The inner city site, previously a vehicle repair garage until 1979, was bought by the municipality at the encouragement of local residents who organised a neighbourhood committee to develop an urban renewal plan to enhance the built environment. The intent was to create more local green spaces, provide opportunities for food growing, play areas, and a public space for social interaction within a densely populated urban neighbourhood. Through extensive consultation between the residents and the local authority, an agreement for the design and management of the site was reached, whereupon the residents would experimentally be responsible for managing the municipal site and receive funding for its redevelopment as a communal garden.

The residents' association *De Bikkershof* was founded after the official opening of the garden to oversee its coordination by the community, preparing an on-going organisational plan concerning environmental, technical and social management (Garlick et al. 2012). The Municipality of Utrecht has since granted the residents group a permanent right of use over the land. Having operated for over 30 years, the *Bikkershof* thus represents an innovative model in self-management of communal growing spaces by citizens to contribute to the greening of the city. This strategy has been stimulated by a liberal approach of the municipality to participatory urban development.

Le Jardin des Aures, Marseille

'Le Jardin des Aures' is a community garden located in the north of Marseille, France that was developed by residents from the economically deprived Mont D'or neighbourhood in 2000. The surrounding area is within the 15th arrondissement of Marseille which has a rapidly changing demographic due to the arrival of incoming populations, whilst the presence of large housing estates is attributed to social and educational challenges in the north of the city. Responding to this situation, local families identified a need for social diversity and spatial cohesion by accessing green space in order to garden together and enable a range of collective activities. The community garden is intended to promote sustainable land use planning, where 'eco-citizenship' raises wider public awareness of urban ecology and approaches to sustainable development.

The *Aures* garden is 0.2 ha in area and sited on former wasteland between social housing blocks. Land is leased from the City of Marseille through a 90-year agreement that, unlike many community gardens, gives this grassroots initiative a secure tenure. It is managed in a collective and dynamic manner that responds to neighbourhood needs and is a project of the *Accueil & Rencontres* association, created by residents to undertake social actions in the area to promote community cohesion. The association is backed by the Protestant Parish of Marseille North which provides a meeting room and community centre to support services provision. The residents' association also attempt to replicate the collective urban garden model by helping several other local gardening initiatives managed by disadvantaged residents without access to domestic gardening spaces. Social actions at the *Aures* garden support a range of environmental and educational activities, including cooking and nutritional workshops, permaculture training, work with disabled, youth and children's extra-curricular activities, and vegetable box distribution to families in need (Accueil 2016).

In the *Aures* case, participation forms the basis of an education in 'urban citizenship'. Participants are socially empowered on a regular basis through all the actions of the local residents so that the users of the community garden are not consumers of neighbourhood services but providers. By developing the gardening project as a social and environmental utility in a working class neighbourhood, the residents' actions are suggested to be motivated against urban processes that can ghettoise and devalue certain areas of Marseille. The wider objective of this social action can therefore be understood as a bottom-up attempt at increasing social justice in the city; a city that now looks towards social and spatial integration of all of its inhabitants and neighbourhoods.

Overview of Urban Food Growing Case Studies from Southern Europe Cites

Pla BUITS, Barcelona

A number of urban growing spaces for community use have been developed in Barcelona, Spain. In 2013, the local authority launched the *'Pla BUITS'* plan, which transferred 14 'empty spaces' in the city from municipal to citizen management, resulting in an increase of participatory food growing initiatives across the city's 10 districts. In order to raise public awareness in urban food growing and help link these new initiatives to existing ones, the *Aplec d'Agricultura urbana* network was formed. It is coordinated by the non-profit association *Tarpuna* with funding from Barcelona City Council. The network is also responsible for organising *L'Aplec* ('the gathering'), an annual public event to raise awareness and develop interest on the socio-ecological aspects of urban food growing, including tours of community gardens, workshops and exhibitions (L'Aplec 2016).

The broader aim of the municipal *Pla BUITS* plan is to regenerate unused land in Barcelona through public activities of a temporary nature. These activities are promoted by non-profit associations or community groups, and subsidised by the city in exchange for encouraging the participation of local inhabitants in the revival of the built environment (Ajuntament de Barcelona 2016). Significantly, the use of vacant urban spaces is restricted from one year to a maximum of three years, must be of public interest or social value, and their facilities readily dismantled when an alternative use is found by the city.

In terms of urban governance, this municipal plan poses questions over residents' participation in the production and interim management of public spaces in Barcelona. It suggests an innovative model of urban planning and development based on the idea of civic participation in 'temporary urbanism' in the context of urban austerity following economic crisis. New functions of state-civil society interactions are being trialled as a response to a lack of financial resources and to resolve the use of abandoned public spaces, along with the demands for increased citizen participation in urban governance. In this regard, the top-down strategy can also be understood as an attempt to formalise the temporary management of unoccupied public spaces with no designated land use through an open and flexible approach. This strategy contrasts the preparation of a long-term development plan for the city with fixed allocation of spaces for collective use.

Le Serre, Bologna

The city of Bologna, Italy has a long history of socialist politics and practice of extended democracy that has enabled citizen participation in the preservation of the urban environment. Dubbed 'Red Bologna' due to its governance between 1945 and the 1990s by *Partito Comunista Italiano*, the main workers' party in Italy, the city has historically applied a social and spatial policy of *'decentramento'*. This decentralisation strategy, facilitated through neighbourhood assemblies, involved the consultation and modification of urban development plans by an increasing number of citizens, often several thousands (Green 1978).

Bologna has since updated the planning principles of cooperative processes and participatory urbanism under a social-democratic administration, and in 2011 launched the initiative 'CO-Bologna'. The purpose is to develop a prototype for the regeneration of an 'urban commons' as a space of social interaction in collectively owned or shared urban resources between its inhabitants. Civic collaboration, social innovation and the collaborative economy are central components to the coordination of various projects in the redevelopment strategy of Bologna (CO-Bologna 2016). In particular, 'innovation hubs' and 'neighbourhood laboratories' encourage public–private partnerships to experiment in developing creative solutions to neighbourhood agendas such as citizen access to land for food growing through use of everyday public spaces. Among these metropolitan hubs of innovation, the case of '*Le Serre*' ('the Greenhouses') at the Giardini Margherita, a public park in the south-east of the city, is exemplary. Located on a 600 m^2 site and former derelict municipal greenhouse, the project represents an emerging place of social and cultural value, acting for the public interest. It is an example of urban revitalisation where community food growing activities are used as a platform for wider entrepreneurship and innovation. The objective is to provide a communal facility and multifunctional space that mobilises local inhabitants in trialling a sustainable and collaborative model of green space governance and to produce new public services. *Le Serre* comprises a community garden, restaurant, day-care centre, and co-working spaces for start-ups and enterprises that is managed on behalf of the city by the Kilowatt cooperative company (Kilowatt 2016).

Le Serre's novel enterprise creates the opportunity for Bologna's innovators, residents, and social economy to work together in responding to neighbourhood needs through an inclusive social, entrepreneurial, and educational model. In terms of flexibility and inclusivity in urban planning, the city is seen to be adaptive as it maintains an experimental strategy of spatial transformation. At the same time Bologna is remodelling a long-consolidated tradition of civic collaboration in its commitment to the idea of 'the city as a commons' (see, e.g. Harvey 2012). Moreover, this municipal strategy is creating the opportunity for social innovations to emerge in urban experimentation. The approach is based on the principles of participatory democracy, involving the collaboration of local stakeholders to contribute to a shared dialogue around urban regeneration.

Terras de Cascais, Cascais

The city of Cascais is a municipality of over 200,000 inhabitants within the Lisbon metropolitan area, located on the Portuguese Riviera. This coastal area is a popular tourist resort and the municipality ranks among the wealthiest in Portugal, though equally has one of the highest living costs in the country (Jornal Economico 2017). In the years following the global financial crisis, Portugal was one of the European countries, along with Greece and Ireland that was particularly affected, and a prolonged recession led to widespread urban austerity measures (Cambon 2008).

In 2009, the Municipality of Cascais began to promote food growing in the city, with the development of its first community gardens. *'Terras de Cascais'* ('Lands of Cascais') is a cultivation initiative that aims to develop innovative green spaces to address a variety of citizen needs, including communal sites for organic growing, revitalisation of agricultural spaces, protecting land access, and self-sufficiency to help reduce monthly living expenses. This top-down initiative can be suggested as, on the one hand, a response to citizen requests for converting municipal spaces into food production sites, whilst on the other hand, a land use policy that could increase public participation while endorsing new forms of urban governance. The holistic strategy adopted by the municipality aims to involve as many participants as possible. It is open to all residents of Cascais and comprises numerous food growing projects, including: Community Gardens; Community Orchards and Olive Groves; Community Vineyards; Associative Gardens (managed by local associations who can sell produce); Gardens in Elderly Day Care Centres, School Gardens; Home Gardens (in private residences); Prison Gardens (training partnership with the Prison of Tires to train and empower prisoners with new skills to enhance opportunities for employment); and a Land Bank (platform between those looking for land to cultivate and landowners) (Cascais Ambiente 2018).

By 2020, over 550 local participants were growing food in 26 new community gardens that had developed 0.6 ha of productive areas in the city (Miguel and Valente 2020). In facilitating this participative approach, municipal land for food growing is introduced in parks and green areas around the city and allocated to residents free of charge. Residents are then responsible for the cooperative management and maintenance of common areas, whilst support and guidance are provided by the municipal gardening team through organic horticulture workshops, meetings, and training. In this sense, mutual help and knowledge sharing between participants is fundamental for maintaining the quality and experience of these spaces. *Ambiente Cascais*, Municipal Environment Department and project administrator for Lands of Cascais, map the various food growing projects around the city and promote and communicate the initiative to interested residents (see Fig. 5.3).

As a result, the City of Cascais, through the distribution and use of municipal land for community growing, is observed to be expanding the development of urban gardening spaces that promote biodiversity, inclusivity, and community resilience through a public service that is integrated into the landscape. By centralising this strategic initiative, the Lands of Cascais represents an ambitious attempt in urban transformation, one that is connected to nature and food growing under a cohesive project identity. Furthermore, the development process has embodied a wide range of cultivation projects, people, and spaces where central to this strategic intervention is civic cooperation in the co-management of food growing areas.

Fig. 5.3 Cascais's urban food growing areas and projects of Lands of Cascais (*Source* Cascais Ambiente 2018)

Overview of Urban Food Growing Case Studies from Eastern and Central Europe Cites

Green Belt and Orbital Forest, Tirana

In Tirana, Albania a strategic level approach to the large-scale development of urban green infrastructure is apparent in the 'Green City Action Plan of Tirana', presented in 2018. The Plan sets out the proposed actions and policy measures of the municipality in tackling ecological challenges and to prioritise expenditure on infrastructures contributing to sustainable development. Contextually, the city experienced a rapid population increase from 280,000 to 860,000 inhabitants between 1989 and 2018. Urbanisation occurred at a faster rate than the development of infrastructure required to sustain this growth and the construction of buildings took precedent over public and green spaces leading to an uneven distribution. The result is a congested built environment with a low ratio of public open space in relation to the number of inhabitants and is worsened by high levels of air pollution (City of Tirana 2018).

A key planning initiative of the City of Tirana is to increase the provision of green space and address urban sprawl through the planting of an 'Orbital Forest'. This project will involve the planting of two million fruit and ornamental trees by 2030, encircling Tirana, acting as a green belt. The Tirana 'Green Belt' aims to connect around 14,000 ha of continuous green spaces in order to improve air quality and maintain biodiversity, and comprises parks, orchards, and other agricultural activities. During the first planting season, over 120,000 trees were planted, many with the participation of citizens and community associations, mobilising public and private land to increase the surface area (APR Tirana 2018).

Tirana's cultivation initiative is innovative in terms of its social value as it increases citizen access to green corridor areas and recreational opportunities while preventing the fragmentation of green space through a productive urban landscape. By multiplying green areas, infrastructure and public spaces, the Municipality of Tirana has shown a political will to create a more sustainable urban environment to protect spatial cohesion and improve the living conditions of citizens.

KZ Vidimova, Prague

'KZ Vidimova' is a community garden located in Prague, Czech Republic, within the mainly residential District 11 to the southeast of the city and is coordinated by the *Kokoza* non-governmental organisation. This gardening initiative is run as a social enterprise and promotes community food growing in the city among residents, companies, and other organisations. *Kokoza* was founded in 2010 by two social entrepreneurs with backgrounds in social work having been inspired by their experiences of visiting similar cultivation projects in England and Germany that combine socially beneficial activities with business. The enterprise currently maps and supports community gardening initiatives in the Czech Republic. The Vidimova garden was established in 2013 as their second community garden following experimentation with a smaller garden in *KC Zahrada*, developed in 2012 as one of the first community gardens in Prague. Contextually, between the mid-1990s and 2013 the city experienced a reduction in half of its 'colony gardens' (former municipal food growing spaces developed during its socialist period) that were lost due to commercial redevelopment following the formation of the Czech Republic (Gibas et al. 2013). Consequently, community gardens have emerged in Prague as civic spaces where novel ideas about urban ecology mutually support gardeners' participation in civil society (Speck 2015).

Vidimova community garden is situated on municipal land and surrounded by tower blocks and housing estates, and as the site is hard surfaced, the 1,870 m^2 food growing areas use raised mobile planters. Around 100 residents are currently involved in cultivation activities and the garden is used as a base to host workshops, seminars and other events that are intended for the general public (Kokoza 2016). A key project aim is the social integration of vulnerable groups into the community, including people who have experienced mental illness or personal addiction, through their employment and engagement in gardening activities to gain an opportunity to participate in everyday life and develop neighbourhood relations. The garden is coordinated by a paid employee of *Kokoza* who performs a significant role as an intermediary between the organisation and members of the community garden as collective stakeholders, in addition to arranging its workshops and public events through a participative approach.

Kokoza as a non-governmental organisation and social enterprise, is observed to have identified the need to demonstrate community food growing and provide educational workshops, having found a vacant area of urban land, and then negotiated with the municipality to lease the land as an external base for a token amount of money. The Vidimova community garden as a spatial product of co-operative social action therefore represents a 'bottom-linked' approach to sustainable development, an approach that is instigated by an urban stakeholder who is external to the local community. This form of urban gardening initiative also suggests a degree of instability concerning the use of municipal land, where agreements are at risk due to potential changes in political representation and local agendas. Furthermore, the long-term sustainability of the garden is currently dependent on the future decisions of *Kokoza*, the approved coordinator of Vidimova community garden. As a result, this means that local residents may need to eventually assume overall responsibility for managing the garden and organising its community activities without support from its present coordinator in order to continue urban development processes.

Onkraj gradbišča, Ljubljana

'Onkraj gradbišča' ('Beyond a Construction Site') is a community-based food growing intervention that has been transforming an abandoned urban space in Ljubljana, Slovenia as a collective social space which encourages residents to participate in planning, organising, and co-managing the city. Comparable to the Berlin case (see *'Prinzessinnengarten'*), this bottom-up attempt at urban planning was introduced in a centrally-located district of Ljubljana on a 'temporary use' basis, and its site is leased from the city on a year-to-year agreement. Launched in 2010, the gardening space within the Tabor neighbourhood is coordinated by the non-profit Obrat Culture and Art Association, a grassroots collaboration between artists (as project initiators), residents and other interested parties. The association grew through developing relations with community members as intended beneficiaries of the intervention through neighbourhood consultations. A variety of social, educational, and cultural opportunities are facilitated in a community space used regularly by around 80 participants for growing food and hosting various public activities (Fig. 5.4) (KUD Obrat 2016).

The spatial organisation of the 1,000 m^2 site in Tabor is continuously evolving, being user driven in its planning, design, and development. In addition to raising public awareness of green spaces in the city and encouraging urban food growing,

Fig. 5.4 Ljubljana urban wasteland at Tabor neighbourhood before and after the development of *Onkraj gradbišča* (*Source* KUD Obrat 2016, CC BY-SA 3.0)

this community space is suggested to be demonstrating an alternative approach to urban revitalisation where space and spatial production forms the basis of social action. Knowledge creation through community action, i.e. 'learning by doing', is supported through reflection in group meetings, exhibitions and in communications with other urban growing initiatives, both locally and internationally.

In terms of Ljubljana's wider urbanisation process, this gardening project is positioned as a citizen response to the reduction of urban food spaces available for public use in the city. Between the mid-1990s and 2008 the area of land allocated for food growing reduced from a peak of 267–130 ha as the city underwent large-scale redevelopment following Slovenian independence (Jamnik et al. 2009). As such, the case suggests a need for municipal support of civil initiatives to expand community food growing. Locally, increasing processes of shared decision-making over neighbourhood spaces contributes to socio-ecological development in Ljubljana. By transferring management of this city-owned space to the civil society through an agreement of land use at zero cost, the City of Ljubljana has demonstrated support for increased citizen participation in the maintenance and self-organisation of urban green spaces.

Overview of Urban Food Growing Case Studies from Northern Europe Cites

Grow Your City, Fredericia

In Denmark, urban community gardens are a relatively new phenomenon in comparison with its European neighbours, such as in Germany or the Netherlands. Yet within the past 20 years, novel forms of cultivation spaces have emerged. In particular, a new type of urban food garden is appearing, one that is formally integrated into large construction projects, initially as temporary spaces then often permanent through relocation, as found in the port-city of Fredericia.

Named after King Frederick III, the Fredericia municipality lies within the greater East Jutland metropolitan area of 1.2 million inhabitants. On Fredericia's seafront, the Kanalbyen Masterplan is a 21 ha development strategy for a former industrial site and shipyard. Begun in 2012, its construction phases will last 20–25 years and when fully developed, this new urban district will accommodate 1,200 homes, and spaces for retail, culture and green areas (KCAP 2016). Overseeing development is FredericiaC, a public–private partnership between Fredericia Municipality and Realdania Urban Development, who have created temporary projects on-site to promote culture and attract interest in the development. Among these interim use projects, 'Grow Your City', is a temporary garden that is testing community food growing and its capability as a component within novel strategies to urban planning. It aims to promote new ways of thinking about construction and urban development; an approach where sustainability, ecology, self-sufficiency, and citizen participation are key agendas. In this regard, the vision draws attention to renewed attempts at supporting nature in cities and spatial activities that can promote more sustainable lifestyles whilst improving the physical and mental health of citizens.

Fredericia's community garden has 600 mobile planter boxes for citizen use to allow for relocation (planned 2022 onwards) to a permanent location at Kanalbyen's nature park (Grow Your City 2016). The garden is used to demonstrate growing and cooking food and uncommon to a community garden, permits residents to cultivate seafood and process fish caught nearby. This 'temporary garden' also acts as a test bed to understand the potential for urban greening and cultivation within the coastal area and its climatic conditions. Successful horticultural and agricultural systems will then be integrated elsewhere around Kanalbyen during development phases.

The Fredericia case is significant as it shows how open spaces, both temporary and permanent, for community food growing are made accessible by urban redevelopment processes. For property developers, temporary projects represent a sound investment, as they create interest in development for minimal cost. These projects can also be utilised to market and promote social and ecological aspects in relation to sustainable urban development. In addition, by demonstrating an ongoing commitment to integrate cultivation spaces within their developments, particularly in the transition of a temporary to a permanent garden, property developers such as FredericiaC can help to sustain citizen access to urban food growing in the long-term.

Odlingsnätverket Seved, Malmö

The city of Malmö, Sweden's third largest city, has an increasing population with over 330,000 citizens (SCB 2018a). Following a process of de-industrialisation in the 1970s and 1980s the city sought to attract new inhabitants with grand redevelopment plans, emphasising sustainability, community development, and cultural diversity. Malmö has since gained a reputation for innovation around urban renewal, though equally the city is socially challenged in terms of segregation and unemployment. For instance, between 2011 and 2013 around 35% of its population originated outside of the European Union and Nordic Countries, whilst it had a higher than average unemployment rate of 9% (SCB 2018b). Around this time, community gardening initiatives with social objectives started to emerge in the city.

Between 2009 and 2011, in the Malmö central district of Södra Sofielund, the *'Barn i stan'* ('Children in the city') gardening initiative worked to integrate Somali immigrant children into Swedish society. The project was developed by a non-governmental organisation in the Seved neighbourhood, a socially deprived area with high rates of crime, unemployment, and social exclusion (Odlingsnätverket Seved 2015). Sharing a similar social goal to the Marseille case (see *'Le Jardin des Aures'*), this initiative aimed to use community food growing as a means to promote social diversity and integration in the area by getting people of different ages and from all backgrounds to grow food together. Following its completion, the *'Odlingsnätverket Seved'* ('Cultivation Network Seved') was formed in April 2011 as a non-profit association in recognition of the contextual need for an organisation that would continue to encourage and support community food growing around housing blocks in the area. The Cultivation Network Seved acts as an intermediary between the municipal owned housing associations, the local authority, and residents. This partnership model means that land and tools are provided by the municipality while the cultivation network's role is to coordinate and educate residents in organic food growing. Key to the network approach is the idea of 'socio-ecological cultivation', where through gardening meetings an exchange in knowledge occurs and social relations are formed between participants. The project's organisation between different local stakeholders further emphasises the utility of employing a central coordinator for community activities and gardening spaces. This coordination links stakeholders and supports project development (see Prague case *'KZ Vidimova'* for similar observation).

Under the support of the Seved network, the small-scale food growing areas around Seved neighbourhood amount to 800 m^2 in area, comprising 30 spaces in public squares, residential courtyards, and a demonstration green wall in addition to a conference room used for meetings. In 2015, around 100 residents were active users of the gardening spaces, of which 60 are members of the 300 strong network (Odlingsnätverket Seved 2015). Due to its popularity, the gardening initiative has since been replicated with the municipality's consent to the nearby neighbourhood of Annelund, Malmö, also managed by Seved network. In spite of the fact that Malmö, and particularly the Seved neighbourhood, has suffered from a negative image due to social exclusion, unemployment and crime, the case shows how localised challenges are producing the conditions for associations and community initiatives to be developed in the city. A process demonstrated by the emergence and replication of community gardening projects, including Cultivation Network Seved.

Reykjavik Municipal Plan 2010–2030, Reykjavik

In Reykjavik, Iceland the city integrated urban food growing within the municipal development plan to promote local food production as a key component. The Reykjavik Municipal Plan 2010–2030, approved by the City Council in 2014, contains a specific environmental policy that for the first time encourages civic spaces for food production be established across the city's 10 districts. Faced with a projected rise from 125,000 to 150,000 residents by 2030 (City of Reykjavík 2014), creating a need for additional housing in the city, the Reykjavik Municipal Plan is novel in emphasising the densification and greening of the existing urban areas in contrast to previous plans which concentrated on the construction of suburban areas. In this strategic level approach, the objective is for a dense, mixed, and diverse urban area that encourages a more sustainable lifestyle in the city's neighbourhoods. Similar to the Fredericia case (see 'Grow Your City'), the provision of opportunities for residents to grow food in the city is a key component to the sustainable design of neighbourhoods within urban master planning.

A recent example of Reykjavik's integration of urban development and food growing is seen in a planning proposal for the city's Laugardalur district (2017—ongoing). A mixed-use 1.45 ha development is proposed on the site of a former municipal energy plant, combining housing, offices, retail, and green spaces with community food growing areas to promote the idea of 'sustainable agriculture'. To further cultivation opportunities across seasons throughout the year, indoor vertical growing and roof-top gardens are also integrated into the spatial design (Fig. 5.5).

In recognition of the contribution towards a healthier and greener city in Reykjavik, the Laugardalur project by Basalt Architects was awarded First Prize in the C40 Reinventing Cities global competition for innovative and resilient urban projects (Reinventing Cities 2020). Overall, the City of Reykjavik, through its innovative

Fig. 5.5 Reykjavik urban planning and development proposal in the Laugardalur district. (Image courtesy Basalt Architects)

approach to urban planning and the formalisation of cultivation practices, shows how policy-based innovation can create opportunities to enhance citizen health and wellbeing and increase biodiversity in sustainable development.

5.2 Conclusions: Social Innovation in Urban Food Spaces in European Cities

In general, social innovation strategies for the production of community food growing spaces in European cities appears to be important in addressing citizen needs and promoting sustainable development. I have shown that this socially oriented approach to urban planning is fundamentally linked to beneficial social and environmental outcomes. The different approaches within each city to innovation and spatial production are typically dependent on the availability of urban land and stakeholder support, whilst directed towards the local needs and goals of community garden users in addition to strategic level objectives. A variety of urban stakeholders are observed to be socially innovating in the development of cultivation spaces. This activity might involve local residents seeking greater public access to green space, social entrepreneurs engaging food growing in their enterprises, non-governmental organisations coordinating community projects, urban developers integrating cultivation spaces within development, or municipal authorities promoting sustainable development.

Based on the city cases examined, the production of community food growing spaces appears to offer particular openings to local inhabitants in civic participation and social empowerment (in terms of urban decision-making and capability building), especially to those who are disadvantaged in some way or excluded from the spaces of everyday life. These environmental processes and social dynamics can occur both internally and externally to public service provision, often in hybrid stakeholder arrangements and represent novel attempts at urban transformations in developing more sustainable lifestyles and environments.

Relating participatory forms of urban governance to the issues of social and spatial justice, the social production of spaces for food growing differs by comparison in the cities studied. For instance, in some cases, such as Berlin or Ljubljana, community gardens are identified by urban activists as political spaces to campaign for greater social justice, where the production of public green space is seen as a way to resist capitalist development and reclaim urban social space. Whereas, in Utrecht, Marseille, Barcelona or Cascais, the use of vacant and neglected municipal spaces for community food growing is more directed towards encouraging and supporting community participation in the revitalisation and regeneration of the urban fabric. Rather than urban food growing being understood here as an embattled concept, citizens in these cases are seen to be given the opportunity to enhance their local environment. Furthermore, citizens are able to develop skills and knowledge and participate in urban governance with the endorsement and backing of the city.

In common to all of the European city cases, the potential for reproduction of social innovation processes in the development of spaces for food growing is observed as dependent on an active citizenship. Particularly, in the mobilisation of engaged local inhabitants who identify the need for opportunities which can increase participation in decision-making processes over the use of public urban spaces.

References

Accueil (2016) Jardin des Aures. https://accueiletrencontres.wixsite.com/accueil/jardin-des-aures. Accessed 28 Jan 2016

Ajuntament de Barcelona (2016) Pla BUITS. https://ajuntament.barcelona.cat/ecologiaurbana/ca. Accessed 10 Feb 2016

APR Tirana (2018) Donate a tree. https://aprtirana.al/une-dhuroj/dhuro-nje-peme/. Accessed 11 Feb 2018

Bewonersvereniging Bikkershof (2016) De Bikkershof. https://www.bikkershof.nl/historie/. Accessed 4 Feb 2016

Cambon D (2008) Budget, taxes, retirement: the lesson of austerity from Portugal. https://www.lefigaro.fr/economie/. Accessed 24 Feb 2018
Cascais Ambiente (2018) Terras de Cascais. https://ambiente.cascais.pt/pt/terrasdecascais/terras-cascais. Accessed 5 Feb 2018
City of Reykjavík (2014) Reykjavík plan 2010–2030. City of Reykjavík, Reykjavík
City of Tirana (2018) Tirana Green City action plan. City of Tirana, Tirana
CO-Bologna (2016) Collaborare è Bologna. https://www.co-bologna.it/chi-siamo/. Accessed 16 Feb 2016
Garlick N, den Hollander F, Figee T et al (2012) 25 jaar Bikkershof: 1987–2012. Bewonersvereniging Bikkershof, Utrecht
Gibas P, Matějovská L, Novak A et al (2013) Garden settlements: shadows of the past or ashes of the future? Charles University, Prague
Green D (1978) What does Red Bologna mean for Britain? Marx Today 195(June):195–198
Grow Your City (2016) Grow Your City. https://kanalbyen.dk/oplev-kanalbyen/byrum/grow-your-city/. Accessed 3 Feb 2016
Hartmann K (2019) Harvest ingratitude. In: der Freitag. https://www.freitag.de/autoren/der-freitag/ernte-undank. Accessed 25 Jun 2019
Harvey D (2012) Rebel cities: from the right to the city to the urban revolution. Verso Books, London
Jamnik B, Smrekar A, Vrščа B (2009) Gardening in Ljubljana. ZRC Publishing House, Ljubljana
Jornal Economico (2017) Lisbon, Cascais and Sintra are the municipalities that best fit with IMI. https://jornaleconomico.sapo.pt. Accessed 24 Feb 2018
KCAP (2016) Kanalbyen Fredericia. https://www.kcap.eu/en/projects/v/kanalbyen/. Accessed 17 Feb 2016
Kilowatt (2016) Le Serre. https://leserre.kilowatt.bo.it. Accessed 29 Jan 2016
Kokoza (2016) Vidimova community and city garden. https://kokoza.cz/work/komunitni-a-mestska-zahrada-vidimova/. Accessed 2 Feb 2016
KUD Obrat (2016) Onkraj gradbišča. https://onkrajgradbisca.wordpress.com/about/. Accessed 12 Feb 2018
L'Aplec (2016) L'aplec d'agricultura urbana. https://agriculturaurbana.cat/aplec. Accessed 10 Feb 2018
Lefebvre H (1991) The production of space. Blackwell Publishing, Oxford
Miguel A, Valente I (2020) Terras de Cascais—from leisure to local food production. Ind Environ 121(1):18–21
Odlingsnätverket Seved (2015) Odlingsnätverket seved. https://www.odlingsnatverket.se/om-oss/. Accessed 2 Feb 2018
Prinzessinnengarten (2019) Prinzessinnengarten. https://prinzessinnengarten.net. Accessed 18 Jun 2019
Reinventing Cities (2020) Lágmúli Reykjavík, Iceland. https://www.c40reinventingcities.org/. Accessed 8 Feb 2020
SCB (2018a) Malmö Folkmängden. https://www.statistikdatabasen.scb.se/. Accessed 24 Feb 2018a
SCB (2018b) Malmö Statistik. https://www.scb.se/Statistik/. Accessed 24 Feb 2018b
Speck C (2015) Growing together: cross pollinating bio- and community food politics at Czech gardens. In: 114th annual meeting American anthropological association. American Anthropological Association, Denver
Utrecht Natuurlijk (2016) Groen moet je Doen! https://www.gmjd.nl. Accessed 4 Feb 2016

Preface to Chaps. 6–11

In the following chapters I analyse six case studies of social innovation in emerging urban food spaces across two UK cities, namely Brighton & Hove and Portsmouth from the perspective of different levels of urban stakeholders. Each case is assigned to each specific organisational level categorised as: residents, non-governmental organisations, and local authority. Cases are presented as a chronological narrative constructed from event sequences and the use of plot structures to support narration. My analysis is undertaken in four interlinked sections to permit the positioning of social innovation processes to case activities. This examination corresponds to the conceptual framework key processes of: 1) Identification of human needs or societal challenges to sustainable development; 2) Development of social relations in systems or structures; 3) Provision of opportunity for social empowerment; and 4) Reflection of socio-spatial development practice. Processes are analysed according to the data coding and research protocol developed in Chap. 2. Each section of my analysis develops the understanding of processes from the previous section. I summarise interactions between key processes and make observations based on the case. Individual case studies are concluded by categorising the influence of contextual forces on social innovation in collaborative spatial development, and key innovation processes are mapped to case activities. The map drawn is used in my cross-case analysis of maps resulting from all six UK-based case studies (Chaps. 12–13).

Chapter 6
Case Study of Social Innovation in Urban Food Spaces: 'The Bevy Edible Garden', Bevendean Community Co-operative, Brighton & Hove

Abstract From 2012 the Bevendean Community Co-operative operated 'The Bevy Edible Garden', a resident-led social enterprise promoting local food and community development in the city of Brighton & Hove, UK. This neighbourhood space integrates community food growing with a training kitchen and café to form a 'community hub', instigating social action through educating residents to grow, cook and eat a healthy diet. It provides training opportunities to develop youth employment and helps to address social isolation of vulnerable residents, while hosting community events to support neighbourhood cohesion. The bottom-up initiative was prompted by local residents and is endorsed by numerous local institutions, organisations, and community groups. Bevendean Co-operative's twelve volunteer directors all live and work locally. Their membership reaches over 800 shareholders and is one of the largest community co-operatives in the country located on a social housing estate. Contextually, the Moulsecoomb estate surrounding the site was built in the 1920s as a social experiment and 'garden estate' inspired by the Ebenezer Howard Garden City Movement of the early twentieth century, though nowadays it is classified as a deprived area. In the 2015 Index of Multiple Deprivation it was ranked 7,826 out of 32,844 in England (where 1 was the most deprived and 32,844 the least). The findings of this case study show that the mobilisation of local residents as a collective is central to social innovation that emerges from the grassroots interactions of civil society, especially self-organisation in urban development.

Keywords Community garden · Community hub · Community co-operative · Grassroots innovation · Community development · Self-organisation · Civic participation · Social innovation · Brighton & Hove

The Bevy Edible Garden (henceforth The Bevy) is studied here as an architype of neighbourhood development in urban food growing and output of community based social innovation, instigated by Moulsecoomb resident and social entrepreneur Warren Carter.

The following examines the development of The Bevy from 2011 to 2017. The enterprise was advanced from 2012 with a share issue to the local community and through a grant from the Social Investment Business Foundation in 2013.

N. Ardill, *Growing Food in Cities*, Cities and Nature,
https://doi.org/10.1007/978-3-030-98475-5_6

6.1 Identification of Human Needs or Societal Challenges to Sustainable Development

The first key social innovation process concerns being 'prompted by an experience or event or new evidence which brings to light a social need or injustice' (Murray et al. 2010, p. 14).

A central identification activity involves residents recognising the need for a community gathering place. The Bevendean Hotel (public house) on the Bevendean and Moulsecoomb estate was closed by police in May 2010 owing to anti-social behaviour and held a reputation as an unsafe and disreputable space. A discussion was prompted amid shared resident concerns that it would be bought by developers as a buy-to-let investment for student housing rather than remain a social space or 'third place', understood by Oldenburg (2001) as between work and home for the community. Focal stakeholder and Moulsecoomb resident Warren Carter, states: 'people need space. It is no good squashing them up in flats with no gardens, no places to meet. I think the idea of space is really [really] important. People need somewhere to meet. Communities need that' (Int_WC). Local Councillor Maria Caulfield was also against property development, at the time stressing: 'I don't think it [the Bevendean Hotel] would be a good site for student accommodation [...] Hillside is a residential area and isn't at all suitable for that sort of development' (Atwell and Arthurs 2010, para. 4).

A resident campaign group was formed in autumn 2010 to protest the site's loss. The need to safeguard public access to space was focussed by an identified threat to community cohesion. Comparable to the European city case studies of Berlin, Utrecht, and Ljubljana (see Chap. 5), local activists here interpreted a requirement to secure a neighbourhood asset through collective governance. Carter stresses how the residents 'thought the first thing we do is own the site, so it cannot become student flats. It would have become student flats eventually if we had not—if this had not worked' (Int_WC). The site was identified by residents as a quasi-public space for social interaction to support health and wellbeing. Locally, opportunities for social relationships were diminishing, especially among older residents through closure of neighbourhood spaces, including the local shopping precinct, public house, and doctor's surgery.

Resident action was also prompted by neighbourhood 'studentification'. This process causes multifaceted urban transformations where large concentrations of students move into specific neighbourhoods, posing challenges to sustainable development, including urban gentrification (Smith and Holt 2007). Locally, houses of multiple-occupancy (HMO) are highly prevalent due to two universities being within a mile of the estate, making properties attractive to investors. To highlight tensions, 'Family Homes Not HMOs' was another resident' group that campaigned to prevent property development. They demonstrated there were over 800 HMOs in Moulsecoomb and Bevendean (40% of the city's total) and petitioned Brighton & Hove City Council (henceforth the City Council) with 1,295 signatures for greater regulation (see Fig. 6.1).

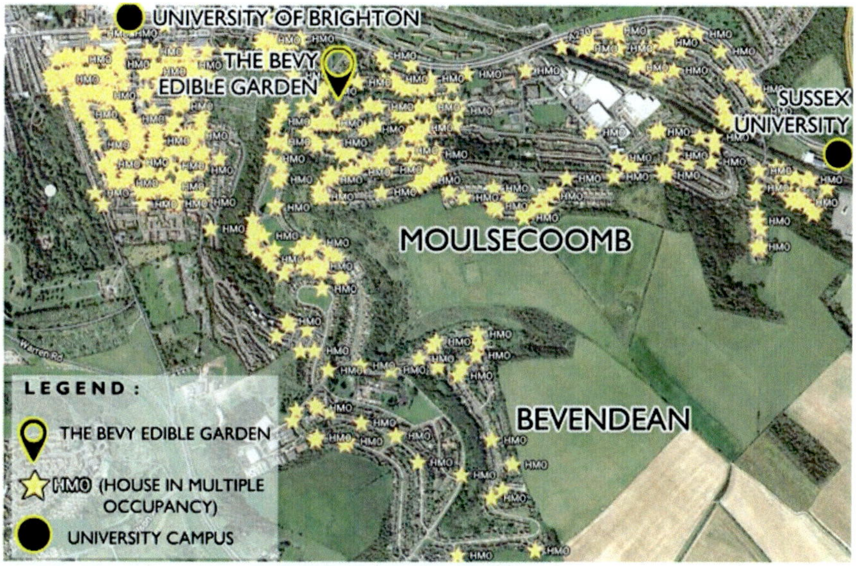

Fig. 6.1 Houses in multiple occupancy (HMOs) on the Moulsecoomb and Bevendean estates (*Source* Author with data from Gandy 2016)

The conversion of commercial properties and social housing into student accommodation has stimulated price rises in the Moulsecoomb neighbourhood through buy-to-let investment (Zoopla 2019). Councillor Anne Meadows, Moulsecoomb and Bevendean Ward, emphasises how 'residents, particularly in Moulsecoomb and Bevendean, are concerned about the impact on their communities. Residents were rightly worried about being priced out of the city' (Duc 2016, para. 52). Discussing this local area dissatisfaction, Carter strongly asserts:

> It has totally f***** the whole of Moulsecoomb and Bevendean, studentification and student accommodation beyond belief. It has mucked up school numbers; it has closed Bevendean GP surgery. It has massively destroyed the infrastructure. Families that have lived here forever cannot afford to live here anymore. You can make more money out of students than you can from people who live here – it has really destroyed the area (Int_WC).

This incoming transient demographic has affected community cohesion through displacement of low-income households. Locally, it has stimulated conflicts with residents, intensified segregation and is understood by residents as accelerating urban decline and loss of public services (Fig. 6.2).

Adversity makes strange bedfellows: the anarchist and the vicar

The Bevy's instigator Carter is a self-identified anarchist, activist and prominent social entrepreneur who previously founded the Moulsecoomb Forest Garden and Wildlife project. The charity specialises in gardening experience for adults with learning disabilities and outdoor education for youths (Int_WC). Additionally, Carter developed an 'Edible Playground' at Moulsecoomb Primary School to embed

Fig. 6.2 Protesters campaigning against the closure of the doctor's surgery in Bevendean (*Source* Duc 2016)

food growing and cooking into the curriculum alongside fellow school governor Father John Wall of St. Andrews Church, Moulsecoomb. Father Wall himself, later becoming a prominent lead and intermediary between stakeholders in The Bevy's development (Int_FJW). Also of significance, is Carter's participation in Power to Change, an organisation supporting community enterprise, and he is a resident director with East Brighton Trust, which supports urban regeneration in deprived areas by using income from property to provide grants. Through the latter association the Trust was prompted to buy Bevendean Hotel in December 2010, with the site later tendered to serve community purpose (Fig. 6.3).

An open meeting was then held in the autumn of 2011 at St. Andrews Church Hall, Moulsecoomb, attended by 25 residents where a steering group was formed, and regular meetings were held thereafter. A doorstep survey of 70 residents was undertaken by a street team to identify local needs and inform strategy for managing the property (Fig. 6.4). This led to a development brief for an 'Edible Garden' and 'Community Hub'; cultivation spaces combined with a community café and training kitchen (see Chap. 5 for a European city comparison with the Bologna case study regarding interlinked spaces, food activities and enterprise). To support community cohesion, collective food production and consumption was central to the aim to develop a 'space for everybody in their own locality to be proud of and to promote a sense of community ownership' (Int_FJW). A neighbourhood share issue was proposed to raise the £200,000 required for development. The premise was if Bevendean and Moulsecoomb's 17,472 residents were to each buy a £10 share, the initiative

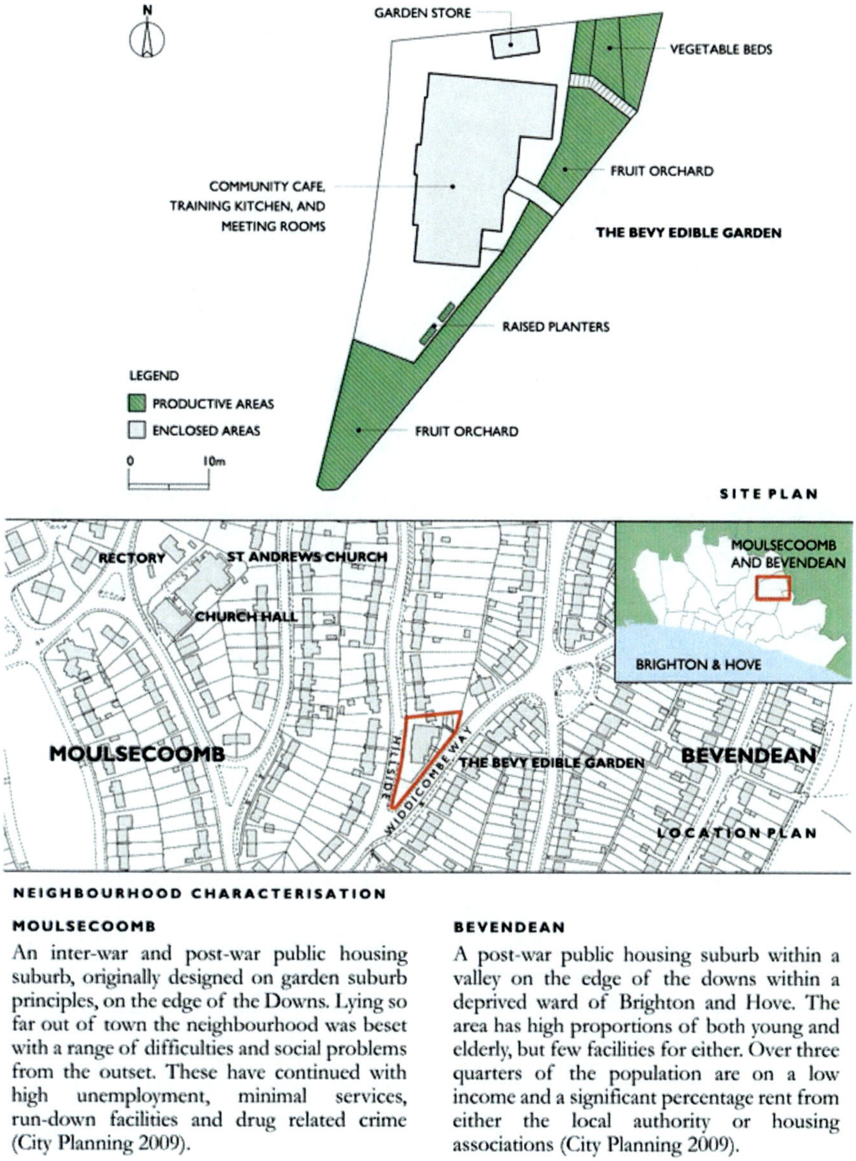

Fig. 6.3 The Bevy site and location plan (*Source* Author)

In the 2015 Index of Multiple Deprivation (IMD) the Brighton & Hove 009B Lower Layer Super Output Area (LSOA) where the site is located was ranked 7,826 out of 32,844 in England, where 1 was the most deprived and 32,844 the least. Indicated by income; employment; health; education, barriers to services; living environment; and crime (DCLG 2015).

Fig. 6.4 The Bevendean Hotel (prior to closure) and mobilisation of Moulsecoomb residents led by Warren Carter, The Bevy's instigator (*Source* Myers and Loman 2015)

would raise the majority of the funding (United Kingdom Census 2011). An affordable figure of £10 per share helped residents on low-incomes and was intended to expand community support through increased membership.

Interpreting the need for 'self-organisation'

The identification of local needs in combination with the retreat of public infrastructure in the area prompted residents to interpret a neighbourhood agenda, become active, and consequently self-organise to undertake community-based development. These residents needed to generate skills and manage resources to support the resilience of their neighbourhood through a process of personal and collective action. In so doing, Carter proposes:

> If you do not believe in what is going on - what are you going to replace it with? If somewhere is going to close, who is going to open it? Governments are not going to. Councils are not. The only people who are going to get it to together and open it are people coming together. If we do not do it - no one else will (Int_WC).

Father Walls adds:

> A lot of the people in Moulsecoomb, they feel well no one is going to do anything for us – therefore we have got to do it for ourselves (Int_FJW).

The local authority helps to corroborate this Moulsecoomb residents' perspective in discussing austerity effects on public services. Emma McDermott, Head of Communities, Equality and Third Sector explains that as a governing institution they have to:

> Start relaying that difficult message carefully to the population to say: we have to share responsibility; we have to do things differently; you cannot expect the City Council to do this anymore. That is a very difficult conversation to have because people are not receptive to it (Int_EMD).

In this sense, the model behaviour of residents as active agents of change, supported local conditions for behaviour imitation and socially innovation. The residents interpreted a need to encourage self-organisation and community driven development, with Carter stressing:

> Organisation. I think you drag people along. We put in a lot of our time for free. I think people can see that and then they volunteer. You set an example. Councils are being shrunk and so the old people's club [Friday Friends] that happens here - there is hardly any left now in this area (Int_WC).

6.2 Development of Social Relations in Systems or Structures

The second key social innovation process involves the active developing of relationships and interaction between two or more individuals, groups, or organisations. This process create the conditions for the exchange of ideas, behaviours, and actions across multiple levels to contribute to the rearrangement and restructuring of existing social relationships (Mumford 2002; Moulaert et al. 2005).

'Intermediation' between urban stakeholders

The influence of innovation agents creating physical and social spaces to mobilise citizens and develop social relations within the case is significant. A key 'intermediation space' was St. Andrews Church Rectory, Moulsecoomb where the local vicar connected citywide stakeholders to generate support for the initiative (Fig. 6.5). Father Wall explains:

> I had this local community leaders' lunch group that I ran for 10 years. It was hugely useful on a number of levels, a whole variety of people coming. The local community enjoyed coming to the house. Because it was a neutral space and I am a good cook and people enjoyed their food, the hospitality, I would have anything between 15 and 30 different community leaders coming (Int_FJW).

Fig. 6.5 St. Andrews Church and Hall, Moulsecoomb and St. Andrews Church Rectory (*Source* Author)

These leaders' lunches at the Rectory brought together influential stakeholders from different social circles over a sustained period. The Rectory facilitated networking and mutual understanding between parties to create relationships that would support innovation. This 'neutral environment' supported an openness of exchange and engagement where space is understood to be experienced the same by different individuals who in turn consider themselves to be equal in terms of power and status (Brown and Baer 2011). Moreover, a levelling process of empowerment and disempowerment is implied to be occurring at the same time within the case, and this neutrality helped to reduce intensity and potential for hostility amongst local stakeholders. The Rectory as a space created for communicating, testing, and advancing ideas supported the development of innovation. Father Wall recalls how:

> Each month someone would do a presentation about what they were doing. People would listen and ask questions […] the number of projects that have actually started; it opened-up a whole number of opportunities. It was hugely [hugely] useful because it gave me a real entree into the life of the town and that part of Brighton (Int_FJW).

The intermediation space permitted an innovation intermediary to integrate into the neighbourhood as an embedded resource, whilst enabling others to connect to different stakeholders and spaces to influence change. Father Wall's agency later extended in scope beyond the neighbourhood:

> One of the local Councillors Anne Meadows became mayor. I became her chaplain. I carried on being chaplain to the mayors; Five or six years. It meant that I could have an entree to the political aspect of the city and see beyond Moulsecoomb. Those were really useful contacts for me. The mayors were great supporters of The Bevy. So, people would come to my leaders' lunch and build up community and that was the heart of what I did in Moulsecoomb (Int_FJW).

Father Wall's relations with city stakeholders permitted access to decision-making in order to generate engagement and top-down support for The Bevy. This increased understanding of the city's political systems and structures, and how to connect neighbourhood issues to citywide agendas, including the health and wellbeing challenges of reduced public services. For example, from 2010 local government spending in Brighton & Hove was reduced by 22%; meaning a change per individual in the city of £490, whereas the national average was £287 (Centre for Cities 2019).

Throughout development, cross-party support and institutional backing was generated for the residents' initiative. For example, Peter West, Mayor of Brighton & Hove, proclaimed:

> In my view The Bevy is probably the best thing going on in Brighton right now […] this cooperative project is a hub for local services and has rapidly become the heart of community life (Power to Change 2016).

Simon Kirby, Conservative Member of Parliament for Brighton Kemptown, further suggested:

> It can play an important role as a hub for the community. I am hopeful this will be the case here in Moulsecoomb. That is why I have been happy to buy a share to show my support for this worthwhile project […] so important that this initiative remains community focused (The Argus 2013a).

In this regard, the community enterprise resonated with the UK Conservative Party's 'Big Society' ideology of the Government (2010–2015) and its emphasis on localism, volunteerism, and the social economy for public service delivery. In demonstrating an active support, Kirby encouraged others to volunteer and negotiated with businesses to get free equipment for the initiative (Int_WC).

The Bevendean Community Co-operative share launch

From the spring of 2012, neighbourhood meetings were held at St. Andrews Church Hall to increase project awareness, recruit volunteers and update progress. A 'Co-operative Friendly Society' status was adopted with support of the Plunkett Foundation (a national organisation specialising in community owned enterprise) and the co-operative would be community owned through membership, whereupon buying shares people become members. Local professionals provided pro bono support during this period in developing a feasibility study for the initiative, including building, financial and legal consultants while a brand identity and logo for the proposed development was designed. In July 2012 a prospectus was prepared to attract investment and additional volunteers ahead of a share launch, and an open day held on-site was attended by over 70 people (see Fig. 6.6).

Dr. Tom Scanlon, Director of Public Health, Brighton & Hove met with Bevendean Community Co-operative (henceforth Bevendean Co-operative) to discuss how the initiative would deliver neighbourhood services, tackle social isolation, and promote behavioural changes in healthier diets and lifestyles (Myers and Loman 2015). Dr. Scanlon later helped the group negotiate local bureaucracy, featured their initiative in the Annual Report of the Director of Public Health (2012–13), and offered endorsement in a promotional film, explaining:

> It is really a community venture. The value of it is, it will give them a meeting place, also give them a bit of identity: we are going to our hub, our community hub, The Bevy. It is a local place, developed by local people where they can meet, cross generations and across social groups to be as one in a community centre (Myers and Loman 2015).

Institutional backing was secured with community participation, social connectedness, and having a say in neighbourhood decision-making being understood as factors underpinning good health (see, e.g. Umberson and Karas Montez 2010). Contextually, the need for action was evidenced by the Moulsecoomb and Bevendean Ward risk of loneliness for those aged 65 and over in the 2011 Index of Multiple Deprivation (IMD) being in the 'High Risk' group and on average ranked 9,890 out of 32,844 in England, where 1 is the highest risk and 32,844 the least (Age UK 2019).

In December 2012 a share launch was held at St. Andrews Church Hall and over 150 residents attended (Fig. 6.7), raising over £2,000 (Myers and Loman 2015). Subsequently, a creative agency assisted Bevendean Co-operative with a 48-hour 'Good for Nothing' hackathon where 'diverse groups support social innovators and changemakers in the places and communities we live' (Good for Nothing 2014, para. 4). Learning was shared through creative commons to support open knowledge advancement. Bevendean Co-operative also worked with designers to configure

Fig. 6.6 The Bevy's top-down endorsement, project identity material and share prospectus (*Source* Images **a** and **b**, The Argus 2013b; **c** and **d**, NEO 2013; **e** to **h**, Myers and Loman 2015)
a Brighton MP Simon Kirby with Warren Carter; **b** Brighton Mayor Peter West at The Bevy; **c** and **d** The Bevy identity concepts; **e** Carter and Father Wall with residents' group at development open day; **f** Dr. Tom Scanlon, Director of Public Health, Brighton & Hove endorsing the initiative in a documentary film; **g** and **h** The Bevy share prospectus

Fig. 6.7 The Bevy's public engagement activities (*Source* Myers and Loman 2015)
a and **b** Co-operative share launch, St. Andrews Church Hall, Moulsecoomb; **c** and **d** Community planting and training events; **e** 'Good for Nothing' innovation hackathon; **f** Local radio interview with Carter and Father Wall; **g** Co-operative share selling stall at local event; **h** Father Wall selling shares to the public in Brighton

development and expand public consultation (e.g. doorstep surveys, leaflet distribution, and online). This activity supported a key grant application to the Social Investment Business Foundation, funded by the UK Department of Communities and Local Government.

In January 2013 the Edible Garden planting commenced with training from the Brighton Permaculture Trust, whilst the event was funded by the 'Harvest Brighton & Hove' programme (see Chap. 7 case 'Racehill Community Orchard', Brighton & Hove). This community event raised public awareness, helped to gain resident feedback, whilst plaques with the names of participants commemorated the occasion. Additional tree planting and training events were held throughout 2013. Fundraising and public engagement activities then intensified to expand community participation and pop-up stalls appeared citywide to increase share selling (Fig. 6.7). Carter additionally wrote in newspapers, gave talks, and was interviewed alongside Father Wall. The local vicar increased media interest, becoming a figurehead, and added to the legitimacy of Bevendean Co-operative.

In late 2013 Bevendean Co-operative was awarded a grant of £130,000 from the Social Investment Business Foundation, the application having been supported by 25 letters of support from stakeholders across the urban system. Owing to their approach to neighbourhood services they were selected out of '210 other organisations because what they could see we are trying to do is unique' (Jones 2014, p. 2). With another £44,000 raised through shares it meant construction could commence. Further aid was given by East Brighton Trust and Social Enterprise Assist investment programme while local businesses donated equipment and tradesmen, designers, and professionals supported development.

Figure 6.8 highlights key stakeholders and relationships in The Bevy development.

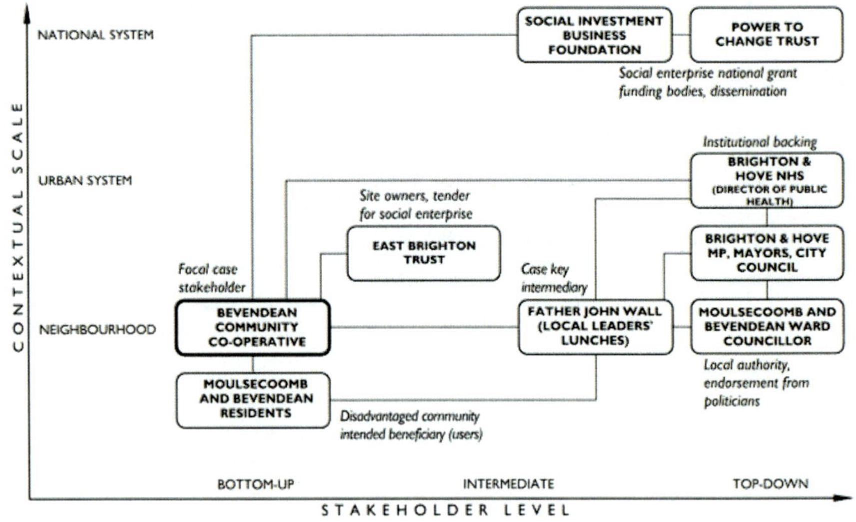

Fig. 6.8 Stakeholders and their relationships in The Bevy development (*Source* Author)
Note Focal case stakeholder outlined in bold

6.3 Provision of Opportunity for Social Empowerment

The third key social innovation process involves the provision of opportunity for collective empowerment through increasing citizen participation and access to power and resources to meet human needs (Moulaert et al. 2005; TEPSIE 2014).

The importance of providing opportunities to the community was evident in this case. Residents sought to affect the situation of the deprived community and local area to support the idea of 'distributive justice'. Carter stresses:

> Working class [people] should have the same opportunities as middle class; we want to offer work experience and opportunities. We want to give kids who may be struggling because of mental health or being in a classroom, outdoor opportunities you know […] The city is very imbalanced through funding and through poverty - it could not give a f*** about the housing estates. It is a little bit scared of them. It does not see any good in them (Int_WC).

Father Walls adds:

> Moulsecoomb was built like a ghetto effectively. It is geographically very much on the edge [of the city] and sort of sealed. It had a sort of sense of being done unto (Int_FJW).

To achieve this social empowerment aim, the initiative acts as a community resource for development, targeting estate youths to improve employment prospects of local people. It offers an alternative classroom whilst increasing openings for confidence building among residents. As an urban green infrastructure, it also increases access to local produce to tackle an area identified as a fresh 'food desert' due to the lack of shops within walking distance (Ridgway 2012).

Increasing community participation

The Bevy's landscaping was completed with the expert guidance of Brighton Permaculture Trust in December 2013. Cultivation spaces total an area of approximately 396 m^2, featuring local heritage fruit trees, vegetable beds, and raised planters. The internal spaces of the community hub were configured by local architects who negotiated statutory consents, ensuring inclusive and accessible spaces that could also serve multiple functions, whilst building work was largely undertaken by volunteers. Overall, the site occupies an area of approximately 1,114 m^2.

In December 2014 the initiative was launched. The event was covered by BBC News South East and used to demonstrate the localised approach to food production and consumption with a social agenda. Students from Brighton Aldridge Community Academy cooked a Christmas dinner for elderly residents in the community café to show how this local space could support community building (the event later becoming a model for a seniors' lunch club).

The training kitchen and community cafe offers a local source of healthy and affordable food in promoting a spade to spoon ethos while surplus from the Edible Garden and interlinked network of neighbourhood food growing spaces is distributed to residents. This network is coordinated by local groups and further connects to citywide organisations to increase volunteering opportunities and distribution of resources (see Figs. 6.9 and 6.11).

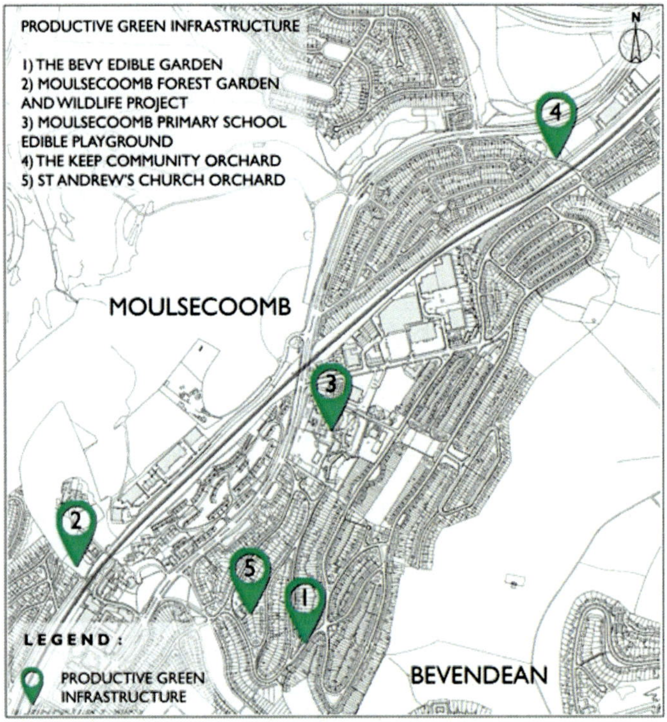

Fig. 6.9 Moulsecoomb's network of productive green infrastructure (*Source* Author)

To illustrate the building of community cohesion, The Bevy hosted a street party attended by over 400 residents in 2016 (Fig. 6.10). An annual garden party and recurring lunch club is attended by up to 50 senior citizens, whilst social events are held throughout the year and internal spaces are personalised with memories and photos of residents (East Brighton Trust 2016). Tackling the social isolation of vulnerable groups locally is thus facilitated through a 'physical and social infrastructure' as Carter points out:

> 70 clubs used the facility last year. A massive mix [...] to attract everyone; the Friday Friends [seniors lunch club] is just a key, resident groups, and all stuff. We want to be a place where I think because The Bevy is here, more things have happened. Be good for our community and the money that we make here stays locally. That is important (Int_WC).

In addition to increasing participation, residents sought to advance social action to benefit the neighbourhood. The social value of goodwill as a generated asset of the development is significant to fostering community cohesion and health and wellbeing through a collective initiative. Here, paid employment supports the local economy and 'people will help out for free because they have a sense of belonging; that sense of being part of is massively important. You need to know your neighbours' (Int_WC). This collectivism highlights the importance of being part of something that people

Fig. 6.10 The Bevy's Edible Garden planting events, training kitchen for local youths and young carers, and cross-generational community activities (*Source* The Bevendean Cooperative 2017)

can believe in and that is bigger than the individual. Snyder and Lopez (2009) for example suggest this quality provides meaning and purpose to support happiness and wellbeing by people feeling more in control and getting more out of what they do. In short, the initiative's volunteers benefit through enhancing their own wellbeing by feeling they are contributing to the community whilst the collective approach promotes community ownership.

By mobilising engaged residents and creating a space for inclusive governance, the economically disadvantaged community participate in collective decision-making processes over neighbourhood space and (as previously stated) with over 800 members The Bevy is one of the largest co-operatives of its type located on a social housing estate (Jones 2017). Moreover, the case suggests a novel approach to resident led development, with Carter pointing out: 'we haven't got retired bank manages, insurance brokers on our management committee. We have retired bus drivers' (Int_WC). Dr. Scanlon further explains this innovative approach:

> Community enterprises are often in rather well-to-do areas, by middle class people who are able to financially invest in the venture - you are not in that situation here. You have a lot of people who are putting in their time, and effort, and a lot of themselves into it. That is the key difference for me. It is very much a bottom-up venture […] I think it really has come to be a model for doing this in other similar [disadvantaged] communities (Myers and Loman 2015).

Annual general meetings are held to enable residents to configure services and collectively vote on decisions affecting the neighbourhood while a friends' group and ambassador roles have dissolved governance and increased community representation. The subgroup liaise with the management committee and gather community feedback on neighbourhood agendas (Jones 2017). Ambassador roles are assigned to several individuals who supported or were personally involved in the initiative.

Numerous services are co-delivered at The Bevy: horticultural training is provided in collaboration with Brighton Permaculture Trust; children are taught how to cook healthy and affordable meals in partnership with the Brighton & Hove Food Partnership (who also ran an experimental dementia café) (Fig. 6.11); NHS Brighton & Hove deliver resident health check-ups on-site; opportunities to develop employability and confidence are provided to people with learning disabilities through partnering with the St Johns College specialist school and Spiral Sussex; the City Council's Learning Disability Employment Team provide work placements and the initiative collaborated with the Widening Access, Research and Mentoring organisation with educational activities for residents experiencing mental health difficulties; the University of Brighton deliver lectures; and the local Councillor holds drop-in surgeries to connect to neighbourhood agendas.

Fig. 6.11 The Bevy's social utility to the Moulsecoomb community (*Source* Photos **a** to **e** and **g**, The Bevendean Cooperative 2017; **f**, Brighton & Hove Food Partnership 2015)
a Friday Friends seniors lunch club; **b** Apple pressing event with children to celebrate national Apple Day; **c** and **d** Distribution of locally grown produce to residents; **e** Annual general meeting for the residents' initiative; **f** Dementia Café; **g** Bevy Ambassador greeting the Lord Lieutenant of Sussex at a garden party

6.4 Reflection of Socio-spatial Development Practice

The fourth key social innovation process involves feedback loops for testing, learning, evaluating and refining ideas, and a range of strategies for growing and spreading an innovation (Murray et al. 2010; SIX 2010; Caulier-Grice et al. 2012; Bund et al. 2015).

The experiences of residents in developing community spaces influenced the project design (Fig. 6.12). For example, in developing an Edible Playground at the local school, Father Wall acknowledges:

> Having learnt from what happened with Moulsecoomb Primary school; that sense of ownership of the community space was something that we very deliberately were trying to emulate […] right from the beginning. There is a sense of public ownership of the school - therefore, it works (Int_FJW).

Locally, Moulsecoomb's school had experienced vandalism and by involving the community in the development of an orchard within the grounds it translated a new meaning as a safe place where people could relax rather than feel threatened (Int_FJW). Likewise, the Bevy faced similar challenges to local resident acceptance due to the negative reputation of the premises. Development knowledge was supplemented through exchanges with a London based group who had converted a public house into a community hub (Int_WC). By building upon experience and replicating others, the Bevendean Co-operative implemented a strategy that promoted collective ownership and shared responsibility to a community space.

Imitation and spreading of resident-led development

The Bevy initiative has won several awards, helping to spread recognition, and a sequence of knowledge dissemination occurred. Father Wall says: 'because it is so high profile, successful, and it seems to be a good model, we had a number of people come and talk to us about it. Members like Warren [Carter] have gone out and talked to them' (Int_FJW). This supported the diffusion of community based social innovation within the urban system and learning was further dispersed to other localities through a network of community enterprises linked to the Power to Change Trust. Carter explains: 'I worked on their Community Business Panel. That was great because you met other community businesses. You learn as much from other people, their ideas and things that worked' (Int_WC). The Trust has promoted a nationwide movement of community businesses to create places addressing local needs. Experimentation throughout development was also evident: 'we were [really] important because we [the enterprise] were so basket case in the first year that we nearly went bust […] because we were the first it is very hard. We are keen to pass on things how we mucked up' (Int_WC). Through trial and error, the initiative was adjusted and refined to evolve the innovation, including site configuration, organisational structure, and service delivery.

The Bevy's visibility communicated a template for development in Brighton & Hove, especially transforming vacant spaces to serve a social utility through community ownership (see Fig. 6.12). Mutual learning was shared with others, for example

Fig. 6.12 The Bevy's dissemination of project learning, local imitation and diffusion (*Source* Photos **a** to **f**, The Bevendean Cooperative 2017; **g** to **h**, Author)
a Power to Change Community Business Panel attended by The Bevy's instigator Carter; **b** Moulsecoomb Primary School Edible Playground; **c** 'Save Exeter Street Hall' resident campaigners with Brighton MP Caroline Lucas; **d** Exeter Street Hall community event; **e** Bevy member Chambers speaking at Meaning Conference; **f** Stoneham Bakehouse, Hove social enterprise; **g** The Lectern former public house in Moulsecoomb; **h** The Open Market, Brighton, run by a Community Interest Company

the 'Save Exeter Street Hall' campaign group in the Prestonville neighbourhood of Brighton where residents sought to buy a church hall to prevent the loss of a community space. Through imitation, they likewise formed a Community Benefit Society under the Co-operative and Community Benefit Societies Act 2014 and issued shares to raise funds and widen participation. Exeter Hall is now resident owned and governed, delivering community services and events. Ian Chambers, local resident, and community activist became involved with both initiatives and shared development knowledge at the Brighton & Hove 'Meaning Conference 2014', an event that networks social entrepreneurs and enterprise to support innovation. Chambers states:

> Galvanising a community to start volunteering and getting involved can be transformational. From my experience of Exeter Street Hall people were saying hello to each other more. We were doing events and meeting each other; it's essentially a community centre […] we have to figure out a way to do things together. If the model is proved, then we can get other community businesses on the [Moulsecoomb] estate: a shop or a laundrette (University of Sussex 2018, para. 4).

In Brighton & Hove, this emerging model of community ownership has continued to develop. For example, Stoneham Bakehouse, Hove is a social enterprise utilising a former bakery to provide educational classes. Other local community groups have likewise approached Bevendean Co-operative with plans to develop properties regarding the threat of housing development to neighbourhood cohesion. Consequently, Carter proposes:

> We are a trailblazer and blueprint about how former pubs can function in the future […] the Greys I spoke to; the Lectern; the Cuthbert people come up; and the Rosehill become a sort of community arts hub. It was going to be housing but they kept it as an arts venue. I do lots of talks, get invited along (Int_WC).

Locally, the spreading of ideas, behaviours and practices across the urban system is suggested to have supported the generation in Brighton & Hove of a networked infrastructure of community spaces, social enterprises, and civil society partnerships.

6.5 Case Discussion and Conclusions

In this section I discuss the interactions between key social innovation processes and summarise my findings. I conclude the following section with a reinterpretation of this case in key processes and spatial scales of analysis.

In summary, I analysed how residents sought to produce a community space to address an identified local need. Residents became project coordinators in the redevelopment of a vacant commercial site in their neighbourhood. Social relations were developed with various stakeholders through cross-sector engagement and supported through intermediation between multiple levels of local participants. The initiative was advanced by a share issue to the local community in combination with a national grant supporting social enterprise. The development of a local facility created social value via collective spatial governance, provision of services, and opportunities for participation whilst contributing to community interaction, cohesion, and social empowerment. Dissemination of project learning took place with others to share knowledge and encourage community-based development.

The identification of local needs centres on the mobilisation of residents to plan neighbourhood development, including the need for self-organisation. By bringing residents together as a group the project stakeholders were able to identify shared concerns and collaborate to find solutions. First, the initiative's founders needed to discover if other local people were interested as the group's structure would not function unless a practical number of residents participated. A steering group helped initially before formalising into a management committee which provided direction and a structure for the group's aims and governance. This approach needed to be open to others to join, be democratic, and keep the community informed about local activity. Two basic approaches to identification activities are evident here. First, 'door knocking', in canvassing (and campaigning) to find out residents' needs and priorities. Second, a series of regular 'open meetings' helped to form relationships in which needs could be interpreted and localised by the community. The former is suggested as restricted, being a one-way communication approach in its potential identification and transfer of knowledge because residents either decline or engage, whereas the latter is more open and can engage residents in ongoing dialogue and exchange of ideas to advance innovation. In particular, the approach encouraged the community to work together to understand local needs, the configuration of the initiative, and the proposed changes to the neighbourhood. Moreover, it established the shared creation of local knowledge in the premise of working with, rather than principally for, the community.

It was shown that across the identification activities, establishing an understanding of the connection between the stakeholders' proposals and local residents was important. This promotion enabled residents to be aware of the opportunity to create a collaborative environment with more engaged and active participants. Participatory planning here enabled individuals to feel they were part of a collective and a valued resource. Moreover, it was necessary for the group to gain legitimacy by engaging with greater numbers of local inhabitants. This enabled an initiative driven by a focal group of residents to be representative of wider opinions and preferences, and consequently neighbourhood needs. The participatory approach extended a shared interpretation of localised condition, experiences, and needs into collective planning, and consequently social action. This protest group was shown to be born out of discontent with external threats, specifically campaigning against property development and the prospective loss of neighbourhood space. However, another significant reason for the group's identification and triggering of community self-organisation as a co-operative was the opportunity presented to develop a community space through collective governance. Self-organisation is therefore promoted by a bottom-up initiative in order to advance neighbourhood development via resident cooperation.

An analysis of how social relations were developed revealed they were influenced by intermediary agents connecting people and ideas, and through creation of ‘intermediation spaces’, both material and social in order to support innovation. This brought together a range of different stakeholders and knowledge required to develop social innovation. The neighbourhood vicar (as case intermediary) enabled necessary bridging, brokering, transactions and knowledge transfer between parties usually separated or restricted from each other. Central to this process was the linking of a group of residents possessing ideas for neighbourhood development with local institutions that have authority and resources, albeit who are potentially less motivated or capable to innovate (i.e. the City Council). This process was shown to be a mutually beneficial relationship between innovation solution provider and seeker. In particular, when residents develop the capacity to affect urban change and local institutions are seen to be facilitating civil society action, especially important to the reputation of local politicians and public attitudes towards local government.

Intermediation between multiple levels of people, groups, and institutions helped to mobilise urban stakeholders. This was observable in activities within and across boundaries that divide sectors with the vicar’s status as mayoral chaplain giving legitimacy to the residents’ group to access political spaces. Father Wall’s knowledge of local politics meant that he was in a good position to translate the needs of the community in connecting to City Council agendas; important to obtaining institutional and cross-party political support for the grassroots initiative. Intermediary agency in supporting a bottom-up approach to social innovation is recognised here in three interlinked stages. First, creating a ‘safe space’ for communication and experimentation as this innovative environment stimulated conditions to instigate social action. Second, in connecting social entrepreneurs with supporters in order to advance their initiative, including influential stakeholders, citywide organisations, and local governance institutions. And third, stimulating the development of local collaborations and social networks to help diffuse ideas and spread social innovation.

Concerning provision of opportunity to local people, my findings indicate the residents managed to meet their objective for collective ownership of an equitable neighbourhood space for social interaction. This approach included ownership over the means of spatial production by the community and for their benefit. With over 800 members, the initiative demonstrates features of ‘collectivism’, in an anti-capitalist struggle over urban development through social entrepreneurism to increase social justice in the neighbourhood. This suggests a shift from the local authority (and subsequent encroachment in private ownership) in the means of producing material space to greater control and management by the estate’s residents as everyday users. Furthermore, it implies that by transferring private assets to common assets held by a group this contributes to the emergence of an ‘urban commons’ (see, e.g. Harvey 2012). Central to this emerging collectivism is the development of cohesiveness among residents and prioritisation of the community interests over the individual. This collectivist view centres on the co-operative’s identity and residents’ common values, principles, and shared objectives. Uniting residents voluntarily through self-governing enterprise and neighbourhood development is therefore important to creating localised conditions for bottom-up, collective action to meet changing sociocultural needs of urban inhabitants.

In terms of the reflection of socio-spatial development practice, it is suggested that imitation is occurring locally due to the influence of bottom-up approaches to social innovation. Other resident groups are observing and replicating the behaviour of the case model for development with social learning helping to establish local practices and cultures, especially through emerging community enterprise and the development of neighbourhood facilities. This process has enabled a grassroots transfer of knowledge in terms of skills, behaviours, and habits between neighbourhood groups. Awareness is therefore a necessary condition to progressing innovation; the enterprise needed to be 'visible' in order to demonstrate and communicate its usefulness to others. Moreover, there is an incentive for others to imitate development practice which in turn causes further reproduction. In short, this sequence suggests a continuous interaction is taking place in which the urban environment is affecting social behaviours which in turn are shaping the form of the urban environment.

6.6 Interpretation of Social Innovation Pattern

I conclude this section by reinterpreting the evolving case in relation to spatial scales of analysis and key processes to develop a pattern of social innovation (Figs. 6.13 and 6.14). The construction of this pattern allows for a summation of contextual influences on case development (see Chap. 2). I will compare the case pattern with other cases in my cross-case analysis in Chap. 12 to identify common patterns to key processes in order to answer my research questions (see Chap. 1).

An examination of the visual map shows a large amount of identification of needs activity within the neighbourhood corresponding to the formation of a campaign group, consultation of residents, and preparation of a development brief for collective enterprise. There follows significant social action to develop relationships across neighbourhood to national scales; creating 'intermediation spaces' and gaining support for a grant application and community share issue on which development was reliant. The social enterprise launch leads to a number of provision activities in the neighbourhood and increased participation through community events.

The case sequence indicates that to produce this community space (and thus provide opportunities) it required earlier identification and development of relations action. Spatial production therefore involved the effective development of social relations to access resources which was dependent on the formulation of an initiative brief identifying the aims and requirements of the social enterprise. A slight variation to my proposed conceptual framework (Chap. 4) in non-linearity is implied with reflection activity occurring both before and after the launch event due to sharing of learning of resident-led development with other community groups and subsequent imitation of practice. My observation does not affect the overall sequence proposed though illustrates how the sharing of knowledge can help to expand urban development and start new cycles of innovation elsewhere.

VISUAL NARRATIVE MAP LEGEND

1. Polygons. The polygons on the visual narrative map describe the main elements of the case chronology relevant to social innovation.

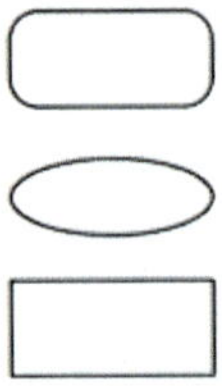

- Rounded rectangles represent activities undertaken by the focal stakeholder
- Ovals represent events undertaken by the focal stakeholder
- Rectangles represent events and decisions outside the control of the focal stakeholder

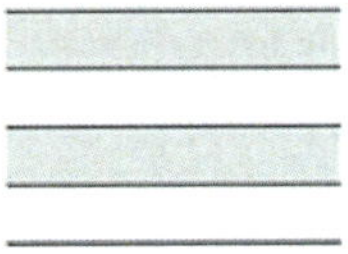

2. Four horizontal stripes. Polygons are situated in relation to the contextual scale in which case activities and events are undertaken. Contextual scales are represented by the ascending horizontal stripes on the visual narrative map of: Neighbourhood, Urban System, and National System. The horizontal stripe to which all boxes are connected contains the base line activity of the core level of focal stakeholders that is indicated in bold.

3. Vertical arrows. Vertical arrows leading from one contextual scale to another indicate direct influences from events and activities on subsequent events and activities.

4. Horizontal arrows. Horizontal arrows within contextual scales indicate continuity of event or activity over time.

5. Dotted arrows. Indicate extended periods of time elapsing within the activity.

6. Time scales. The time scales along the bottom of each visual map vary in scale to minimise the size of maps whilst emphasising significant periods of activity and events.

7. Key social innovation processes. Key processes are represented by colour-coded polygons and situated next to the events and activities they relate to following case analysis according to the study protocols.

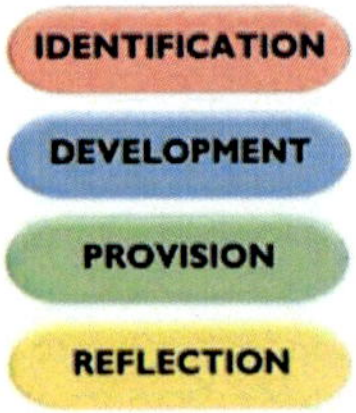

- Identification of human needs or societal challenges to sustainable development
- Development of social relations in systems or structures
- Provision of opportunity for social empowerment
- Reflection of socio-spatial development practice

The visual narrative maps can be read as an overview of the main events and activities within the case narrative and represent the complete case study incorporating narrative subplots. They provide a visual summary of the case study and through signifying key processes a graphical sequence from which sequence analysis can be undertaken.

(*Source* Author)

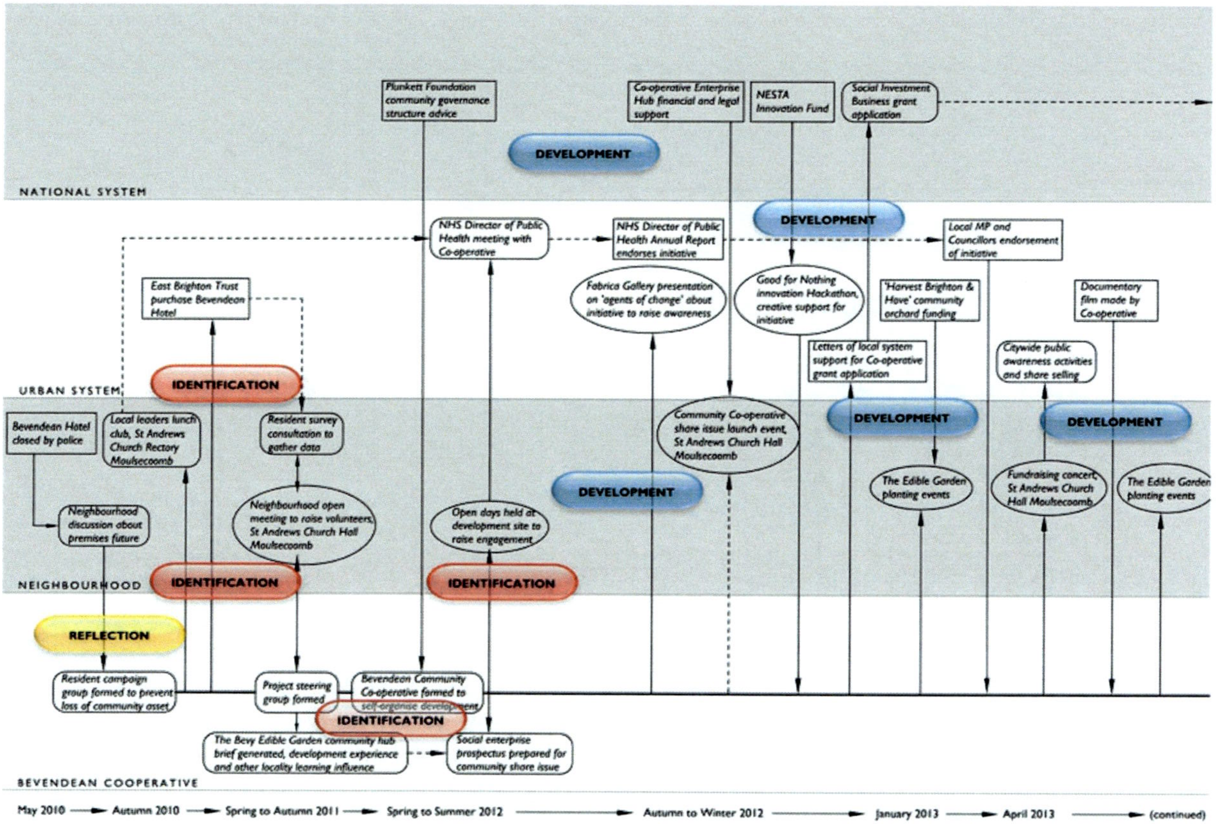

Fig. 6.13 Visual narrative map of The Bevy development with key social innovation processes indicated across contextual scales. Map sheet 1 of 2 (*Source* Author)

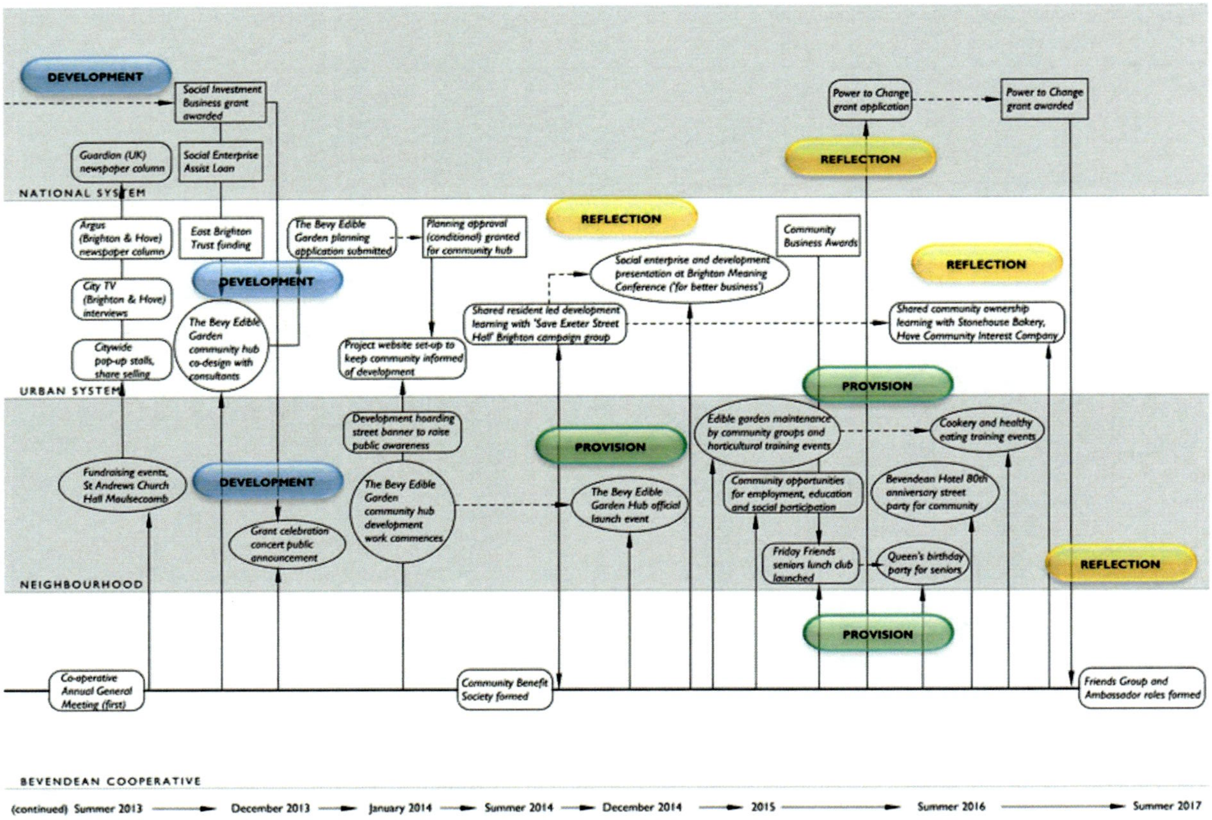

Fig. 6.14 Visual narrative map of The Bevy development with key social innovation processes indicated across contextual scales. Map sheet 2 of 2 (*Source* Author)

Table 6.1 provides a summary of contextual influences across spatial scales on development.

Table 6.1 Summation of contextual influences on The Bevy development

Contextual scales	Influences on project development
National system (Macro scale)	• Social Investment Business Foundation, principal grant, set guidelines • Power to Change, principal grant, set guidelines • Social Enterprise Assist, loan and financial mentoring • Plunkett Foundation, aided setting-up governance structure • Co-operative and Community Benefit Societies Act (2014), influenced governance structure • Learning shared by resident groups in other localities to support planning and development
Urban system (Meso scale)	• East Brighton Trust, purchase site and tendering process, prompted the initiative, set guidelines • Vicar intermediation between urban stakeholders and top-down support from NHS Brighton Director of Public Health, Member of Parliament for Brighton Kemptown, local councillor endorsement • Negotiate planning permission for development with local authority • Citywide public and media engagement events • Good for Nothing ideas collaboration, supported resident consultation and funding approaches • Harvest Brighton & Hove, small grants, built relations, set guidelines • Co-production of services with local organisations and institutions
Neighbourhood (Micro scale)	• Closure of neighbourhood public house, prospective loss of social space, understanding of neighbourhood needs • Perceived sense of social and spatial injustice, threat to community cohesion • Area of multiple deprivation, lack of opportunities for community development • Resident-led campaign group facilitated neighbourhood engagement • Previous experience of green space development by group • Community consultations, identify collective needs • Resident willingness for self-organisation and interest in spatial governance • Enterprise community share issue and membership, community buy-in • Friends group and ambassador roles, increased community representation and devolved spatial governance

Source Author

References

Age UK (2019) Loneliness at local and neighbourhood level. https://www.data.ageuk.org.uk/loneliness-maps/england-2016/brightonandhove/. Accessed 24 Jul 2019

Atwell O, Arthurs M (2010) Proposed Bevendean Hotel sale angers locals. In: The Argus. https://www.theargus.co.uk/news/8423804.proposed-bevendean-hotel-sale-angers-locals/. Accessed 18 Jul 2019

Brighton & Hove Food Partnership (2015) Brighton & Hove Sustainable Food Cities Award 2015. Brighton & Hove Food Partnership, Brighton & Hove

Brown G, Baer M (2011) Location in negotiation: is there a home field advantage? Organ Behav Hum Decis Process 114(2):190–200

Bund E, Gerhard U, Hoelscher M, Mildenberger G (2015) A methodological framework for measuring social innovation. Hist Soc Res 40(3):48–78

Caulier-Grice J, Davies A, Patrick R, Norman W (2012) Defining social innovation. European Commission, DG Research, Brussels

Centre for Cities (2019) Cities outlook 2019 a decade of austerity. Centre for Cities, London

DCLG (2015) National Statistics File 7: all ranks, deciles and scores for the indices of deprivation, and population denominators. Ministry of Housing, Communities & Local Government, London

Duc F le (2016) Brighton campaigners urge councillors to stop family homes from being turned into shared houses. https://www.brightonandhovenews.org/2016/10/21/brighton-campaigners-urge-councillors-to-stop-family-homes-from-being-turned-into-shared-houses/. Accessed 19 Jul 2019

East Brighton Trust (2016) Friday friends at The Bevy. https://www.eastbrightontrust.com/blog/read_165929/friday-friends-at-the-bevy.html. Accessed 17 July 2019

Gandy B (2016) ePetition details Family Homes Not HMOs. https://present.brighton-hove.gov.uk/mgEPetitionDisplay.aspx. Accessed 17 July 2019

Good for Nothing (2014) Good for nothing—how it works. https://www.goodfornothing.com/how-it-works. Accessed 11 Sep 2018

Harvey D (2012) Rebel cities: from the right to the city to the urban revolution. Verso Books, London

Jones H (2014) Bevendean Community Hub AGM Minutes 2014. Bevendean Community Hub, Brighton & Hove

Jones H (2017) Bevendean Community Hub AGM Minutes 2017. Bevendean Community Hub, Brighton & Hove

Moulaert F, Martinelli F, Swyngedouw E, Gonzalez S (2005) Towards alternative model(s) of local innovation. Urban Stud 42(11):1969–1990

Mumford MD (2002) Social innovation: ten cases from Benjamin Franklin. Creat Res J 14(2):253–266

Murray R, Caulier-Grice J, Mulgan G (2010) The open book of social innovation. NESTA, London

Myers B, Loman P (2015) The Bevy. Homegrown Films, Brighton & Hove, UK

NEO (2013) REINVENTING. https://www.weareneo.com/our-thoughts/. Accessed 1 Nov 2017

Oldenburg R (ed) (2001) Celebrating the third place: inspiring stories about the "great good places" at the heart of our communities. Da Capo Press, Chicago

Power to Change (2016) Bevendean Community. www.powertochange.org.uk/what-is-community-business/stories. Accessed 14 Sept 2018

Ridgway T (2012) Brighton's fresh food deserts. In: The Argus. https://www.theargus.co.uk/news/10096854.brightons-fresh-food-deserts/. Accessed 8 Jan 2019

SIX (2010) Study on social innovation. The Young Foundation, London

Smith D, Holt L (2007) Studentification and "apprentice" gentrifiers within Britain's provincial towns and cities: extending the meaning of gentrification. Environ Plan A 39(1):142–161

Snyder CR, Lopez S (eds) (2009) Oxford handbook of positive psychology, 2nd edn. Oxford University Press, New York

TEPSIE (2014) Social innovation theory and research: a summary of the findings from TEPSIE. European Commission, DG Research, Brussels

The Argus (2013a) The Bevy. In: The Argus. http://www.theargus.co.uk/news/10298673.The_Bevy/. Accessed 31 Oct 2017

The Argus (2013b) MP buys a share in Brighton community. In: The Argus. http://www.theargus.co.uk/news/10219450.MP_buys_a_share_in_Brighton_community. Accessed 1 Nov 2017

The Bevendean Cooperative (2017) The Bevendean Cooperative. https://www.facebook.com/pg/TheBevy/photos/?ref=page_internal. Accessed 1 Nov 2017

Umberson D, Karas Montez J (2010) Social relationships and health: a flashpoint for health policy. J Health Soc Behav 51(S):S54–S66

University of Sussex (2018) Iain Chambers. https://www.sussex.ac.uk/about/community/heroes/iain-chambers. Accessed 19 July 2019

Zoopla (2019) House prices in Moulsecoomb Way, Brighton BN2. https://www.zoopla.co.uk/house-prices/east-sussex/brighton/moulsecoomb-way/. Accessed 17 July 2019

Chapter 7
Case Study of Social Innovation in Urban Food Spaces: 'Racehill Community Orchard', Brighton & Hove Food Partnership, Brighton & Hove

Abstract Brighton & Hove Food Partnership was formed in 2003 due to the identified need for a partnership approach to integrate sustainable urban policy, agency, and social change. It emerged as an umbrella non-governmental organisation within the city of Brighton & Hove, UK connecting cross-sector stakeholders to form a participatory and strategic approach to developing a holistic food system. The organisation is now embedded in the city with over 4,000 members and links food policy with initiatives within public health, education, community development, land use, urban planning and sustainable development. From 2009 to 2013, 'Harvest Brighton & Hove' was an innovative citywide programme of the Food Partnership to develop local food projects. Altogether, Harvest supported the development of 54 new growing projects across the city; transforming 1.19 ha of urban land into productive green spaces. As a Harvest exemplar, the Racehill Community Orchard was the most significant community space to be developed with permission to grow to 1.30 ha, and consequently is the largest orchard in the city with potential to produce 4,000 kg of fruit annually. Contextually, the Whitehawk estate which borders the Racehill Orchard was, in 2015, the most deprived area in the city and the 332nd most deprived area in England, placing it just outside the bottom 1%. The findings of this case study show that an intermediary organisation connecting neighbourhood initiatives, the local authority, and community groups was significant to social innovation, especially for improving citizen access to urban resources.

Keywords Community gardens · Community orchards · Urban food growing · Urban development · Innovation intermediary · Food Partnership · Social innovation · Brighton & Hove

Brighton & Hove Food Partnership (henceforth the Food Partnership) participate in local governance partnerships through the City Sustainability Partnership, Local Strategic Partnership and the Health and Wellbeing Board. The organisation coordinates local activities with national agendas through their sister body Food Matters whilst connecting to the UK Sustainable Food Cities Network to share knowledge.

The Racehill Community Orchard (henceforth Racehill Orchard) is studied here as an output of practitioner based social innovation and exemplar of the 'Harvest

N. Ardill, *Growing Food in Cities*, Cities and Nature,
https://doi.org/10.1007/978-3-030-98475-5_7

Brighton & Hove' urban development programme, overseen by Helen Starr-Keddle, Harvest Development Officer.

The following examines Harvest and Racehill Orchard from 2008 through to 2015. The development of urban green infrastructure was advanced from 2009 by the Big Lottery Local Food Fund and Brighton & Hove City Council support with public land to develop for food growing.

7.1 Identification of Human Needs or Societal Challenges to Sustainable Development

A central identification activity is the Food Partnership recognising that 'to address the complex problems in our current food system, whether from a public health or an environmental point of view, is never more pressing than now' (Else 2008, p. 15). This national agenda was highlighted in several Government reports; the Cabinet Office Strategy Unit (2008) review into the UK food policies emphasised challenges to meeting basic universal needs around equity, health, safety, and environment, whilst the Department of Health report 'Choosing Health' (2004) identified local community food initiatives as an instrument to support behavioural changes and reduce health inequalities.

In recognising this situation, the Food Partnership diagnosed challenges to sustainable development with local needs confirmed through collaborating with Brighton & Hove City Council (henceforth City Council) and working with Food Matters (Int_HSK). Relationships were reinforced by Francesca Iliffe, City Council Sustainability Officer holding a seat on the Food Partnership's Board (Food Matters 2011). Iliffe proposes that the partnership approach enables these organisations to 'move easily between community and corporate stakeholders and are a critical friend and ally to the City Council in establishing one of the leading cities in the UK for food work' (Williams 2019, p. 1).

A significant prompt to development was the Local Food Fund, launched in 2007 under the UK Big Lottery Fund's Changing Spaces programme to help communities' access and improve their local environments. It presented an opportunity to confront a national agenda at the urban level with a £59.8 million fund awarded to food related projects to make locally grown food accessible (The Wildlife Trusts 2016). Local Food sought to build capacity across three levels: (1) 'Material capacity' in provision of physical infrastructure; (2) 'Personal capacity' in contributing to personal development and empowerment; and (3) 'Cultural capacity' through engaging with the local community and ensuring their buy-in (Kirwan et al. 2012). This approach was intended to develop the overall capacity and resilience of communities involved. Social innovation was actively positioned in the programme as a major driver of capacity building:

> Underpinning the notion of capacity is the concept of 'social innovation'; it is about encouraging changes in social practice. […] the specific aim of increasing levels of participation,

> especially amongst those who had previously been excluded in some way; in so doing, those involved are empowered to take more control over their lives and to take a more active role in society (Local Food 2012, p. 8).

The Food Partnership's social agency complemented the aims of Local Food. Starr-Keddle, Development Officer explains: '[how] the ideal scenario is that we help a project; they carry on their own and do fantastic work because we are in a sense capacity builder's' (Int_HSK). By enhancing the value of capabilities, citizens are understood to develop the capacity act to bring about changes in their urban situations to affect wellbeing and quality of life (see, e.g. Fainstein 2005, 2010). Locally, the context for developing capabilities through urban food growing was:

> [A lot] to do with the climate at the time; there was a lot of people interested in setting up community food growing spaces and not a lot of training or being able to talk to each other, share knowledge and skills. So that [presented an] opportunity; there was a need out there and no one was filling it. Then the Local Food Fund became available (Int_HSK).

In this sense, Brighton & Hove presented the 'right city in the right time' for this undertaking due to the embeddedness and connectedness of the Food Partnership in combination with local grassroots interest to develop urban food growing spaces.

Translating a national agenda into an urban agenda

The Food Partnership recognised an opportunity for community food initiatives to contribute to meeting a local requirement to reduce inequalities in relation to health. Detailed understanding was gained from working with the local authority to deliver food related health services and developing the City Food Strategy Spade to Spoon: Making the Connections (2006) (Int_FI). This Strategy set objectives to widen resident access to nutritious food and opportunities for urban food growing (Brighton & Hove Food Partnership 2006). Brighton & Hove, although situated in the affluent south east, was found to be within the 25% most deprived local authorities in England owing to 8% of its neighbourhoods being among the poorest 10% in the country and 21% among the most deprived 20% (OCSI 2007). The survey exposed a social equity divide between rich and poor, and this situation was underlined by the city's Annual Report of the Director of Public Health (2006). The Public Health Report stressed challenges of growing health inequalities in child obesity, low adult intake of fruit and vegetables, inactive lifestyles of local inhabitants, and deprived wards having a life expectancy of up to five years below more affluent wards in the city (Brighton and Hove City Primary Care Trust 2006).

The compilation of evidence to highlight local needs was thus a key identification activity to development. A persuasive argument was made for the proposed solution to satisfying needs by evidencing a strong interest in local food, and its impact on the environment and health. It was demonstrated that over 1,500 visitors attended open days at cultivation projects in the year leading up to Harvest (Else 2008), while over 1,200 people were waiting for an allotment in Brighton & Hove—representing one of the longest waiting lists in the country (Department for Communities and Local Government 2006). Iliffe reveals how the city 'had a real problem in terms of lots of people wanting to do food growing but not the sites to accommodate them. So, part of the whole reasoning behind Harvest was to create more areas of land being used for food growing' (Int_FI). Existing research was also incorporated as consultations showed citizens were keen to support local food production and wanted access to locally produced food.

In preparing their Harvest bid document, the Food Partnership organised meetings with members and partners, including the City Council and Primary Care Trust, to jointly develop the proposal's aims, and outcomes, as well as the delivery of initiatives. Explaining the social benefits of this approach, Starr-Keddle says: 'Because we [the Food Partnership] work in partnership; every project we do we deliver with other partners. It adds a lot more strength into our work to be able to work so closely with other people' (Int_HSK). Individual meetings with City Council Officers kept them informed of Harvest and gathered feedback to support an integrated strategy. Mita Patel, Sustainability Programme Officer, suggests the role of Iliffe in connecting the activities of the Food Partnership to the City Council has added value to the city: '[because of] having that middle person between the Council and the Food Partnership promoting the Food Partnership agenda, the work that they do has grown [and grown]' (Int_MP). Furthermore, the Sustainable Cabinet Committee were encouraged to support the Food Partnership because the 'work by the Food Partnership is likely to lead to further collaboration and joint working from which the Council can benefit from support and resources from the Food Partnership' (Iliffe 2009, p. 13).

The collaborative development of Harvest by partners within the urban system representing the identified 'beneficiaries' (i.e. local inhabitants and users of cultivation spaces) was a strategic approach to satisfying needs. Several of these partners had links to the Food Partnership and experience of delivering community food projects. In addition, participatory planning with residents facilitated trust building prior to any delivery work by involving target neighbourhoods in project development. Between June to October 2008, numerous citywide events and public workshops were held and attended by over 130 people (Fig. 7.1). A questionnaire assessed interest in food growing, support to increase food growing activities, and potential obstructions. There were 270 responses; 189 respondents (70%) were interested in participating in a growing project if nearby. This confirmed a demand for more local food, growing food, and becoming involved in a community food growing project (Else 2008). Feedback also helped to tailor Harvest components, including a training programme that responded to the needs indicated by prospective participants (Int_HSK). In short, the sequence of identification in this case demonstrated that cross-sector consultations were influential to interpreting and communicating local needs, and thus suggests an important factor in developing a cohesive strategy for urban green infrastructure.

The Harvest Brighton & Hove bid submitted in November 2008 was innovative in adopting a systems-based citywide approach to urban food growing (see Chap. 5 for comparison with a European city strategic level approach in the Cascais case study). At the time, projects in other cities typically concentrated their agency towards a specific institution, demographic, such as schools and older people, or focussed on a single cultivation project (Int_HSK). By way of contrast, Harvest deliberately operated at multiple levels to link top-down and bottom-up processes. First, to ensure long-term support for the project aims, it worked with the City Council at a strategic level to develop policies and embed good practice. Second, it worked with residents and

Fig. 7.1 Harvest Brighton & Hove public consultation poster (*Source* Brighton & Hove Food Partnership 2008)

neighbourhood groups at the grassroots level to deliver its aims. In terms of administration and coordination, the Food Partnership 'acted as an umbrella to distribute funding to other organisations so that they could do projects' whilst coordinating events and networking parties (Int_FI).

In late 2009 the Food Partnership was awarded the maximum grant of £500,000 from the Local Food Fund. Harvest would act as 'Beacon Project' to share learning to support projects in other cities within the UK and internationally (Kirwan et al. 2012).

Racehill Community Orchard, Whitehawk

Selected food growing projects were designed as Harvest exemplar spaces for developing models in standardised land leases or to demonstrate community food growing on public land. Learning from exemplar spaces would be used to develop protocols and agreements for growing projects between community groups and the City Council (Brighton & Hove Food Partnership 2014).

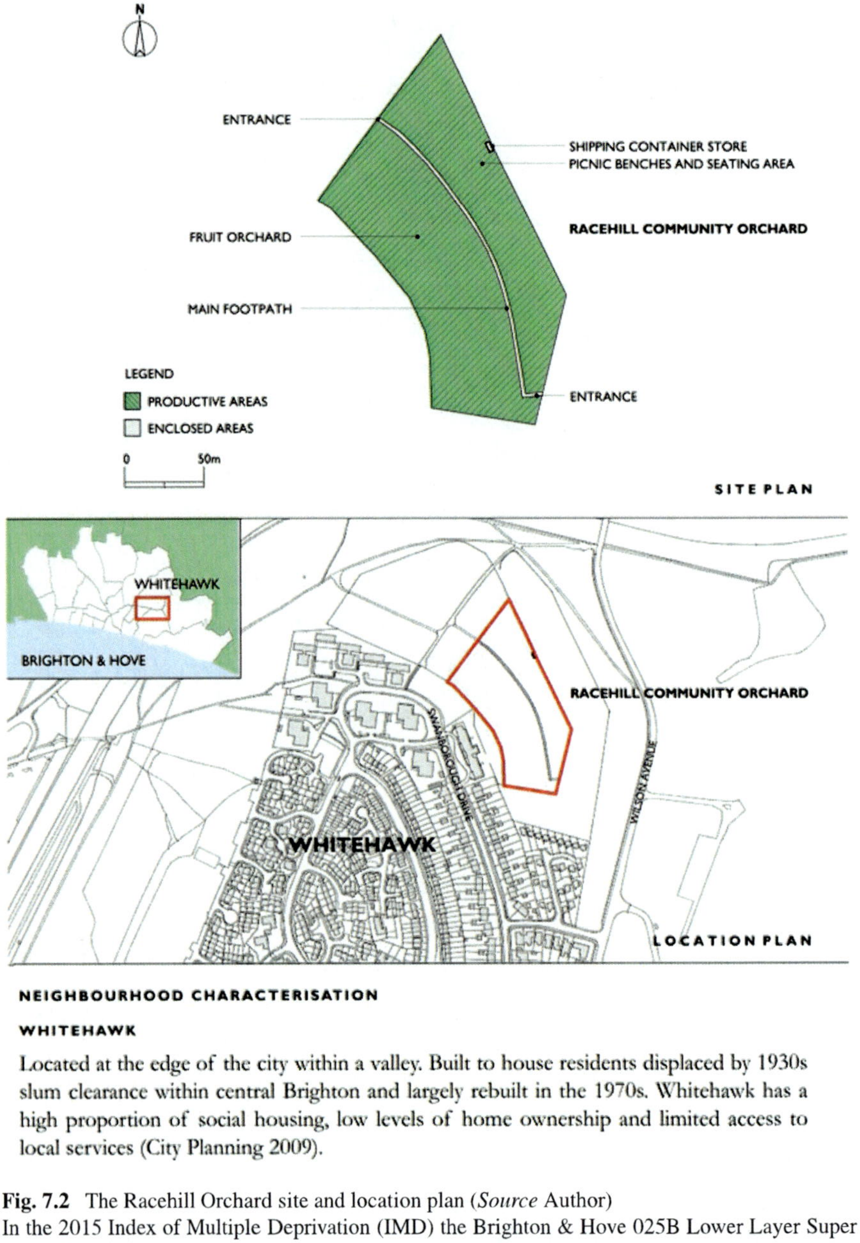

Fig. 7.2 The Racehill Orchard site and location plan (*Source* Author)

In the 2015 Index of Multiple Deprivation (IMD) the Brighton & Hove 025B Lower Layer Super Output Area (LSOA) where the site is located was ranked 332 out of 32,844 in England, where 1 was the most deprived and 32,844 the least (Department for Communities and Local Government 2015).

Harvest aimed to benefit residents citywide whilst also ensuring that areas of socio-economic disadvantage would especially gain from spatial interventions to address poor access to fresh food, high incidence of poor health, and lack of access to urban resources. Evidence was gathered for an exemplar community orchard to be located within the Whitehawk neighbourhood of East Brighton on a site near to social housing tower blocks (see Fig. 7.2). The high-density estate was identified as a location experiencing social needs and an intervention was proposed here to enable more residents to participate in food growing activities (Int_HSK).

To support community engagement, a consultation of Whitehawk residents was undertaken and neighbourhood plans were developed in consultation with the target community. Starr-Keddle explains how 'we [the food Partnership] always do a consultation before anything happens because the worst thing that can happen is that something is done to a community and then they basically get really angry' (Int_HSK). This extensive consultation process helped to substantiate an identified neighbourhood need whilst alleviating any resident concerns. Neighbourhood backing was gained through meetings with local organisations and resident groups, exhibitions, and a questionnaire. The closer people lived the more interest they expressed in the development. 192 respondents (88%) wanted a community orchard. 63% lived in the Whitehawk area, and 85% lived in the BN2 postcode area of East Brighton accessible from the site. Residents stated several reasons for support:

> Transforming an unused local area into a productive space; making the area more attractive; to bring residents together, meet people and be a community asset; support biodiversity in the local area; benefit children; provide an educational and learning experience; give residents better access to local fruit; and save the site from housing development (Starr-Keddle & Borrill 2011, p. 6).

Residents additionally sought community participation, collective powers, and increased opportunities:

> Encouragement to people of Whitehawk specifically to be very involved in any project on that area of land [...] A great opportunity to engage people on low incomes to become involved in a project that offers a chance to be really involved in a local initiative that offers decision-making, and agency to truly local people to that area (Starr-Keddle & Borrill 2011, p. 7).

7.2 Development of Social Relations in Systems or Structures

Collaboration with other Brighton & Hove stakeholders through partnership working was central to the case. This process helped to strengthen and develop new or existing relationships within the urban system and was aided by intermediation between groups across issues and communities. For instance, Racehill Orchard involved a partnership between the Food Partnership and Brighton Permaculture Trust, an organisation which promotes sustainable development through design (Int_BT). The Trust was a Harvest partner and delivered key programme areas, especially orchard planting in the city. Racehill Orchard was developed as a publicly accessible site and

'Demonstration Growing Garden' to attract and introduce residents to food growing (Int_HSK). The development was jointly led by Starr-Keddle (community events) and Bryn Thomas, Brighton Permaculture Trust (horticultural activities). As such, the partners multiplied their practitioner and development agency through sharing resources and expertise to realise a more effective solution than that which would have occurred by working independently. Consequently, this approach opened-up new avenues and means to support social innovation.

City Council support for urban food growing

The City Council provided key support for urban food spaces, with most of Harvest's initiatives located on public land. For Racehill Orchard, Thomas approached the City Council in saying: 'we are a partnership of Brighton Permaculture Trust and the Food Partnership; we are looking for a potential site. That involved a Council officer actively considering sites, identifying them, and suggesting them to us. It was my job to go around and check them out' (Int_BT). City Parks represented an important link here between the City Council, project partners and the community, and needed to be supportive of any new activities in permitting development.

The City Council recognised the potential social value of the Racehill development to the neighbourhood whilst also adhering to its strategic urban policy as part of its commitment to food growing and sustainable development. For example, the Brighton & Hove Sustainable Community Strategy (2006) aimed to increase land available for food growing; the City Food Strategy (2006) sought to increase cultivation opportunities; and the updated Food Strategy Spade to Spoon: Digging Deeper (2012) targeted 'more food consumed in the city is grown, produced and processed locally using methods that protect biodiversity and respect environmental limits' (Brighton & Hove Food Partnership 2012, p. 14). The influence of local politicians also factored. Thomas says: 'we've had a number of points of contact with Gill [Mitchell] around stuff. All the three [city] parties are supportive to the idea of community orchards' (Int_BT). East Brighton Councillor Mitchell adds:

> It's great that we have been able to plant so many orchards. They benefit the community in so many ways and the fact we have planted one which may be the largest in a century in Brighton & Hove is such a brilliant achievement (Brighton & Hove City Council 2016, p. 1).

The Racehill Orchard agreement drawn up between the City Council and project partners was significant because this initial 15-year land agreement enabled support from grant bodies to sustain development, with Thomas stressing:

> We wouldn't get the funding if we didn't have a secure lease because the funding comes to us and that's easily the most persuasive argument when talking with the Council. It's a quick and succinct discussion. We want some commitment from the Council if we are going to invest this much time and energy (Int_BT).

This commitment from the City Council to the project aids the development of productive green spaces as apple trees are planted for a long-term yield and 'it is part of the strategy; the longer you tie it up with a lease the more security you've got to get on and develop it for the long term' (Int_BT). Furthermore, the terms helped to facilitate the development through a 'peppercorn lease' (i.e. a token rent) and the permitted public land use, setting conditions of:

- Free public access to be maintained at all times;
- To benefit the local community through engagement and outreach with local and citywide residents, local tenant associations, community groups and schools. Provide education in fruit tree growing to local residents;
- Produce grown to be distributed within the community, donated to local schools, donated to volunteers or other community organisations;
- Excess produce may be sold through local markets with proceeds re-invested in the project. Sales of juiced fruit grown on the site permitted. To be run as a non-profit organisation (Brighton & Hove City Council 2011).

The implications of this secured agreement meant that it served as a model for shared governance between the City Council and community groups for other Harvest cultivation spaces around the city which were later developed. Racehill Orchard was additionally important in helping to change attitudes to community food growing on public land, such as concerns of vandalism or being visually unattractive, through demonstrating an ongoing commitment to spatial management (Brighton & Hove Food Partnership 2014). This built trust amongst the City Council to support community food growing with access to common resources. Similar to the role of non-governmental organisations seen in the European city cases of Prague and Malmö (see Chap. 5), the Food Partnership is positioned within this case as mediator between stakeholders potentially affected by an urban food growing project (Int_FI). They work with community groups interested in setting up a project and use their influence to liaise with the City Council to gain preliminary consent through an archetypal sequence. First, community consultations help to configure cultivation spaces. Second, if an initiative gets a positive reception, the Food Partnership assists in the preparation of a project plan, setting out land management. And third, submits the plan to the City Council for approval (Int_HSK).

Locally, the implications of developing relationships between city stakeholders have meant that organisations such as Brighton Permaculture Trust now have 'such access to City Parks that we find out more quickly and easily who the person is to deal with. It works in quite a devolved way, so if the local person is in favour it seems to happen quite easily' (Int_BT).

Learning generated by Harvest enabled the creation of a template for community agreements with the City Council that outlines roles and responsibilities for each partner, maintenance arrangements, and conditions for governing areas of public green spaces (Int_HSK). Neighbourhood groups now govern several of these spaces or co-manage with Parks Services, and through deregulating public space this process increases citizens 'right to the city', especially in the formation of an 'urban commons' as proposed by Harvey (2012). As such, communities are now actively encouraged to participate in the production of green spaces amid public service austerity. In discussing the approach of the local authority, Emma McDermott, Head of Communities, Equality & Third Sector suggests:

> What we are trying to do is to have a different conversation: do you want to do it? We are not going to stand in your way [...] my team are moving barriers out of communities' way. I can regale you with tree planting. So, we can't give you any money, we can't give you any officer anymore or very little but what we could do is give you the right to if you like (Int_EMD).

Furthermore, the need for the City Council to adapt to emerging governance structures is understood as:

> Those are the conversations we are beginning to have through Local Area Action Teams and Friends of Parks groups. All those types of things that are beginning to shift that cultural understanding of us, but also the culture within ourselves about how we work with residents (Int_EMD).

The community template has also informed approaches in other cities. For instance, Sustain (the alliance for better food and farming) used learning from Brighton & Hove to share experiences of managing food growing projects in public settings to publish a local authority guide (Sustain 2018a).

The 'co-production of space' at Racehill Orchard

A cross-sector advisory committee for the Racehill Orchard aided project partners in organising the development and ensured that interested parties were democratically represented. The participatory framework enabled a co-design process to promote a sense of community ownership and helped to collectively develop a spatial and programmatic brief. An ongoing management plan set out governance structures. Thomas says: 'the original idea was that the place would be handed over in terms of day to day management, though we [Brighton Permaculture Trust] would remain responsible to a local group' (Int_BT). Subsequently, resident volunteers formed the Racehill Community Orchard Group and supported tree planting in late 2010. The intent was to enable a direct line of communication between the City Council as landowners and residents as custodians. To activate local people, the creation of intermediation spaces in the neighbourhood influenced relationship building, as Thomas explains:

> It has taken quite a bit of networking to get known in the local community. There was one of these [tower] blocks had a community meeting room which then closed down, but we met there, and residents who live near here [Racehill Orchard] we met in their house (Int_BT).

These 'brokering spaces' helped to engage and configure a community of volunteers whilst local knowledge and connections generated additional support from local groups to accelerate development. The ongoing governance of Racehill Orchard was proposed to be eventually transferred to Whitehawk residents. However, the opportunity presented to the community to gain direct control over urban space, and thus fulfil 'radical' interpretations of social innovation as a driver of structural changes in power relations (see, e.g. Moulaert et al. 2005), failed to be assumed. Rather, the residents decided:

> We would rather this was run as a subset of Brighton Permaculture Trust [...] those involved made a decision that they didn't want that additional level of work, they didn't want to be an independent organisation (Int_BT).

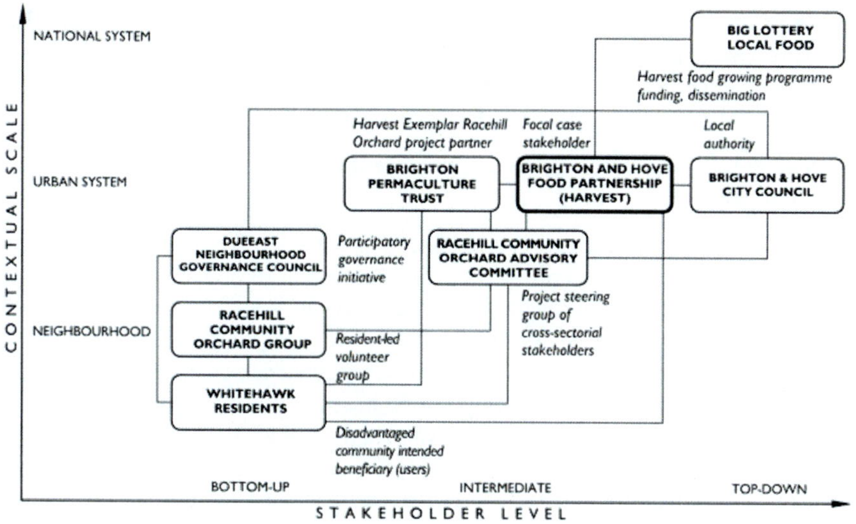

Fig. 7.3 Stakeholders and their relationships in the Racehill Orchard development (*Source* Author) *Note* Focal case stakeholder outlined in bold

Circumstances were further compounded by the later withdrawal of several residents following initial enthusiasm. Thomas laments how 'we've been a bit unlucky with the individuals from the local community that we did manage to attract, being part of making the place what it is, but not being involved longer term' (Int_BT). As such, the case demonstrated a desire among residents to remain in externally coordinated governance structures which could also help to continue initiatives.

Figure 7.3 highlights key stakeholders and their relationships in the Racehill Orchard development.

7.3 Provision of Opportunity for Social Empowerment

In this case material capacity was increased by developing physical infrastructure and improving public access to green space; these factors were complemented by educational opportunities that built personal capacity. Harvest facilitated community development through training workshops, skills-sharing and open days which were delivered citywide to residents, often in community food growing spaces; thereby supporting engagement. The programme increased capacity through building confidence in cultivation and enhancing capabilities for managing projects (Int_HSK). Overall 65 courses were delivered on cultivation-related topics to over 550 attendees, 97% of attendees felt more confident after training and 92% said they planned to share their learning with someone else (White et al. 2013). Harvest linked emerging projects to existing sources of support to access resources and ran a mentoring programme, matching up experienced projects with others needing support to start-up or to scale-up. This development of physical and social infrastructure in Brighton & Hove was supported by the Food Partnership being a central point of contact and coordination (Int_HSK).

Racehill Orchard: building capacity, widening participation, and emerging public space

At Racehill Orchard, residents from Whitehawk estate were trained in agricultural techniques and offered 'activities and visits, our volunteer sessions; we get groups up here to visit and try to engage with local residents in terms of taking an active role in making decisions or getting involved with the site' (Int_BT). This process helps to build personal capacity. Thomas says: 'in terms of feedback we get from our participants - they feel enabled, skilled, empowered to get on with stuff afterwards. That's definitely one of our drivers and our motivations' (Int_BT). The social value of these opportunities enhances communities' capabilities, while volunteers can be trained in leadership skills through a session leaders' course; helping to devolve organisation and share knowledge (see Fig. 7.4). This approach to social behavioural change supports community empowerment through enhancing neighbourhood capacity to act, improves resident access to skills and resources, and helps cultivation projects to potentially become self-sustaining.

Several free activities and recurring community events were delivered to Whitehawk residents at Racehill Orchard. To demonstrate user participation and community engagement, from June 2014 to May 2015, two large public events were attended

Fig. 7.4 Racehill Orchard training events and Healthy Activity Days (*Source* Brighton & Hove Food Partnership 2015a)

by 134 people from the Whitehawk area postcode (BN2 5). Regular 'Healthy Activity Days' were also held, involving health walks, tree planting, foraging events, pick and cook sessions, and other activities which promote healthy lifestyles and behaviour. Overall, 163 adults attended 560 times and three volunteers were trained in leadership skills (Brighton & Hove Food Partnership 2015a).

Material capacity was strengthened in Whitehawk by increasing the accessibility of green space to local residents. The site was previously a wasteland as Starr-Keddle Points out: 'you could barely walk through; quite painful and difficult to navigate. Now every time I go there, there is different people using the site, loads of people are now using the site that would never have done before' (Int_HSK). Thomas adds: '[Racehill Orchard is] like a public park to a lot of people. People come up here and sit on the benches, enjoy the sunshine, walk their dogs. It is definitely a public space, and we are developing it along that vision, it's within the residents' ability to decide' (Int_BT).

The production of green infrastructure in bringing new land into food production, providing community events to build capacity, and improving the access to resources of specific target groups is reinforced by enabling public access and rights to urban space. In this sense, the observed 'right to the city' being supported in this case connects to the arguments made by Lefebvre (1991) and Harvey (2009) for spatial justice and increased citizens' rights over urban spaces. Moreover, cultural capacity was developed here through public engagement and territorial appropriation to give residents a sense of connection with their urban landscape. This process is significant to the 'social production of space', where space operates as both a product and a producer of changes in the urban environment (see, e.g. Lefebvre 1991; Soja 2010).

The importance of the observed socio-spatial dynamics to Racehill Orchard means that 'where people have ownership over a space then they are more likely to look after it. At Racehill we have got several local people that walk through the site every day - they really want to look after it' (Int_HSK). This spatial appropriation helps to prevent any defacement whilst reinforcing local responsibility. Furthermore, 'it gives them [residents] pride. There are flats that overlooks the orchard, so they are physically seeing something that they might be directly involved in, that gives a sense of belonging and a different quality of appreciation for that space' (Int_FI). To support place attachment, local youths from the Princes Trust at Whitehawk Academy painted a mural at the site to identify it as their neighbourhood space, whilst a picnic area was created with memorial benches dedicated to local people who helped to develop the initiative (Fig. 7.5).

In order to maintain public services, access to neighbourhood resources, and tackle inequalities (and thus continue social innovation) funding was necessary, especially after Harvest ceased in 2013. The deprived urban condition was influential to grant funding, with Thomas stressing:

> The Whitehawk estate next to us - this top end is the most deprived area on the south coast. That undoubtedly has helped us to attract funding for working in this area. We have received quite a bit of funding from different sources; it is about engaging the local community and very much about targeting disadvantaged communities (Int_BT).

Fig. 7.5 Racehill Orchard site photographs (*Source* Author)

The awareness of third sector support, in combination with the long-term site lease and public access were also key factors in obtaining several grants: £25,000 was obtained from Veolia Environmental Trust; £10,500 from the Big Tree Plant Government-sponsored campaign; and People's Health Trust funding (Brighton Permaculture Trust 2013a). East Brighton Trust also funded some benches and free foraging and cookery workshops (East Brighton Trust 2016).

Racehill Orchard's identity, biodiversity and contribution to urban ecology

In order to brand the Racehill Orchard identity, a logo was created and used for online communication and on information boards' onsite that provided information to visitors on fruit trees, plants and wildlife (Fig. 7.6). Additionally, in terms of 'rewilding' urban spaces, the initiative connects to a grand societal challenge concerning biodiversity. Research demonstrates that between 1970 and 2015, 58% of animal and plant species declined, and the UK has lost significantly more nature over the long term than the global average. Transformative change is considered essential for the nation to meet the 2030 Agenda for Sustainable Development and Sustainable Development Goals (Hayhow et al. 2019). Racehill Orchard features a managed wildflower meadow to increase biodiversity, over 230 fruit trees and 2,250 native fruiting bushes to conserve vulnerable local fruit varieties, while being grown in organic conditions to promote ecological principles (Brighton Permaculture Trust 2013b).

Fig. 7.6 Racehill Orchard identity material, ecology, and public engagement (*Source* Graphics, Brighton Permaculture Trust 2015; Photo, Author)

7.4 Reflection of Socio-spatial Development Practice

A prominent case activity was the diffusion of ideas and behaviours by amplifying the visibility of social practices. Harvest aimed to get more people in Brighton & Hove growing food by increasing accessible land and people participating to help reduce health inequalities. However, realising this aim is challenging as a geographically-constrained city between the South Downs National Park (40% of the city area lies within) and the English Channel. It is a highly populated and highly developed urban area with limited available space within the city (Brighton & Hove City Council 2014). Yet in terms of developing material and cultural capacity, Brighton & Hove's community food growing spaces tripled from 25 to 79, with many projects located in new areas, including housing estates, public parks, church yards, railway stations, and universities (Fig. 7.7).

Harvest's integrated citywide approach helped diffuse social innovation and generated an urban infrastructure in social relations, networks, and resources. The programme has helped new people participate in urban food growing through training and events with this practice subsequently involving over 4,000 volunteers per year (Growing Health 2015). The distribution of Brighton & Hove's cultivation spaces before and after Harvest demonstrates how access to new spaces has increased

Fig. 7.7 Brighton & Hove's community food growing spaces developed by Harvest (*Source* Author)

(see Figs. 7.8 and 7.9). This is especially evident among deprived areas to support 'distributive justice' in the socially just allocation of resources and to increase the public visibility of growing practices. The Food Partnership and Food Matters have since mapped areas of land around Brighton & Hove that might additionally be used for cultivation (Int_FI).

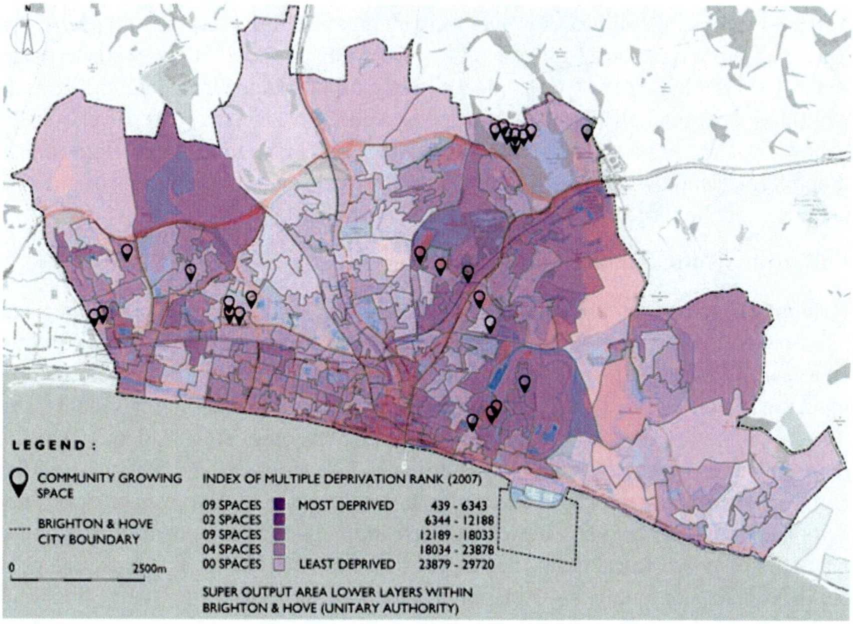

Fig. 7.8 Brighton & Hove's 25 community food growing spaces in 2009 (before Harvest) mapped against areas of deprivation (*Source* Author with data from Starr-Keddle 2017 and Department for Communities and Local Government 2007)

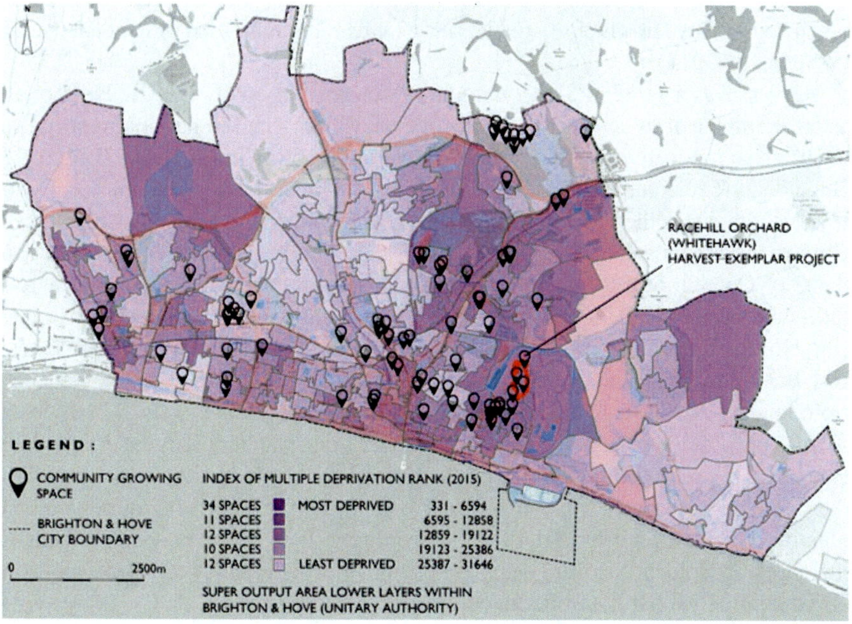

Fig. 7.9 Brighton & Hove's 79 community food growing spaces in 2015 (after Harvest) mapped against areas of deprivation (*Source* Author with data from Starr-Keddle 2017 and Department for Communities and Local Government 2015)

There were some failures however in other areas of Harvest. For example, 'Grow Your Neighbour's Own', a residential garden share initiative was unsuccessful because available gardens were typically on the urban fringe while potential growers without space generally lived centrally so needed to be willing to travel (further confirming the need for distributed growing spaces). The initiative experienced high dropout rates and owing to the administration support required it was recommended it would operate more efficiently at a neighbourhood scale (Int_HSK).

Project dissemination and shared learning

A central objective of the Harvest experiment as a Beacon Project for the Local Food programme was to share experience and knowledge of the project's citywide approach to increasing food production, and to disseminate learning that could be replicated in other cities, both within the United Kingdom and internationally. This was undertaken in several ways. A Reference Group comprising academics, project representatives from other urban food projects, policymakers and planners enabled parties to learn from the experiences of nationwide projects. Harvest activities were demonstrated to visiting city representatives and the Food Partnership 'visited other cities to share our learning and learn from others' (Int_HSK). They also attended pan-European conferences on planning and food systems to exchange thinking. Local Food facilitated project dissemination through Share and Learn networking events and a national evaluation event (Fig. 7.10). Additionally, annual reports and a final evaluation report were distributed to influence local and national policymakers through demonstrating outputs. Information was openly shared for other places to access, whilst press articles and articles in specialist publications increased coverage. Brighton & Hove was also a founding city partner in the Big Dig (Good to Grow) organised by Sustain which shared learning to over 20 urban food projects in the UK (White et al. 2013).

Harvest was awarded Best Community Growing Project by the UK Big Lottery as an exemplar of good local food practice and highlighted in national reports and media as an innovative model for an integrated urban growing strategy (Brighton & Hove Food Partnership 2015b). Locally, Racehill Orchard as an exemplar project of Harvest influenced the further development of green infrastructure across Brighton & Hove to increase resident access to food growing. The project partners 'worked with the City Council and the local community of Hollingbury and created a map of the estate where all the green spaces were and then planted trees over that estate' (Int_HSK). Additional orchard planting was undertaken at the Craven Vale Estate, East Brighton to regenerate public spaces in deprived areas, with 67 trees planted there from January 2012 over a two-year period with the help of residents (Fig. 7.10). Starr-Keddle proposes 'because we have got such a good reputation now and they [City Council] have seen the work done in other places, that actually they are keen to have these food growing sites in the city' (Int_HSK). In Spring 2021, community orchard planting was extended by Brighton Permaculture Trust to several new locations, including at Coldean, Hollingdean, Bates, and Bevendean Estates in the north of Brighton & Hove (Brighton Permaculture Trust 2021).

Fig. 7.10 Harvest programme influence and dissemination of learning (*Source* Photos **a** to **b** and **e** to **f** Author; **c** to **d** and **h** Brighton & Hove Food Partnership 2015b; **g** Local Food 2012)
a and **b** Craven Vale Estate Orchard, Brighton & Hove; **c** and **d** Craven Vale Estate community planting; **e** and **f** Hollingbury Orchard, Brighton & Hove; **g** Local Food 'More than just the veg' national evaluation event; **h** Sustainable Food Cities UK networking event

In this case the distribution of social innovation knowledge was supported by the innovativeness of the organisation, as shown by the Food Partnership mediating between different project stakeholders, leading to Brighton & Hove becoming a leading urban voice on developing community food projects. Their agency helped to mobilise a social movement of community growers and develop a network of cultivation spaces, whilst engaging local decision-makers helped to influence planning and development policy in addition to attracting the attention of national policy makers (see Chap. 5 Barcelona case study for a similar European city example in the development of a local food growing network). In helping to drive food policy in other cities, the Food Partnership undertook a learning exchange on urban food growing and planning with Sustain, who used this experience to meet with London planners to discuss adopting planning guidance to support food growing within development plans. This experience was further shared with other cities and organisations (Int_HSK). Building on this knowledge, Sustain published the report 'Planning sustainable cities for community food growing' encouraging city leaders to integrate community food growing within policy to meet strategic objectives (Sustain 2014).

Following Harvest, in 2013, 'the Sustainable Food Cities Network was set up based on some of our experience here and by sharing what we've done it could be extended' says Iliffe (Int_FI). The UK-wide network shares knowledge of urban experimentation and innovation in local food agendas. Brighton & Hove is held as a model of good practice for other localities through its citywide strategic approach to development (Durrant 2015). The network enables localities to be connected to wider national campaigns and empowers residents, campaigners, and local authorities to make local policy changes through a collective voice. Starr-Keddle explains: 'we try to influence policy and one of my targets supporting the local community is to help the community feel like they could influence policy and vice versa - so it is helping them adapt to change' (Int_HSK).

Nationally, the Food Partnership was approached to give oral evidence on Harvest to the House of Commons Environmental Audit Committee on Sustainable Food in 2011. This insight focussed on planning approaches to support local food systems and sustainable development. The Food Partnership highlighted how partnership working enabled them to unite different sectors working in the local food system to amplify Harvest and other community food projects, and how this approach was aided by Brighton & Hove being a geographically compact city. The stimulus of a top-down strategic approach to food in having a national food strategy was proposed as a mechanism to address other local authorities not also taking strategic approaches (Environmental Audit Committee 2012). This agenda item continued with Michael Gove, Secretary of State for Department for Environment, Food and Rural Affairs (Defra) proposing a National Food Strategy in 2018, the first such nationwide policy in the UK since World War II (Sustain 2018b). In July 2021, The National Food Strategy: The Plan was published with a strategic objective to 'Make the best use of our land' and a key recommendation to 'Invest £1 billion in innovation to create a better food system' (Dimbleby 2021). Significantly, the review recognises the intrinsic value of a local food strategy, developed and delivered in partnership with communities, to bring about the necessary food system change.

7.5 Case Discussion and Conclusions

In this section I discuss the interactions between key social innovation processes and summarise my findings. I conclude the following section with a reinterpretation of this case in key processes and spatial scales of analysis.

In summary, I analysed how a non-governmental food organisation through the action of practitioner based social innovation increased food growing practices, and thereby addressed an identified local need to increase the amount of land available and the number of participants. The Food Partnership built on existing social relations whilst further developing multilevel relations between community groups and the local authority and utilised national grant funding to become project coordinators of an urban food growing programme, advancing the development of productive green spaces. Social value was created through building community capacity to enable residents to obtain, improve, and retain skills and knowledge of food growing. Opportunities for community participation were widened, especially among disadvantaged communities. Lastly, engagement activities occurred in multiple ways, both within the UK and internationally to distribute urban development knowledge.

The identification of needs was shown to be influenced by the organisation's partnership approach. Their cross-sector membership added insight into local needs through gathering data from different sources to compile evidence for potential solutions. Central to this agency was extensive consultation with residents, delivery, and institutional partners. From the outset, Harvest was planned and designed in collaboration with partners, rather than through an ad-hoc approach. A systematic, purposeful approach to planning is observed in creating conditions for effective project delivery. This is due to the fact that Harvest connected the identification of participant's needs and understanding of local challenges to the satisfaction of the service user needs. Harvest also helped to ensure that all partners had adequate procedures, structure, and capacity in skills, experience, and resources to advance development at the neighbourhood level.

My findings suggest three steps of identification activity by Harvest's coordinators. First, internal consultations with members and partners. This involved meeting with individuals, communities, and statutory organisations to collectively develop objectives, outcomes, and delivery of Harvest components. Second, organising and attending workshops and public events as part of the consultation process to inform detailed development. Third, use of questionnaires to assess community interest, and to gauge enablers and barriers to urban food growing. These three steps helped the Food Partnership and project partners to develop and refine procedures, specifically by responding to the information proposed beneficiaries gave through consultations as to what they needed. This process contextualised urban to neighbourhood needs and is explained by a citywide agency, drawing together a coalition of diverse partners, and the combination of strategic level and community-based agency in the gathering and generation of new knowledge.

On examining how social relations were developed, this was shown to be supported through a process of coordination and connecting. As the focal point for information, the Food Partnership organised multiple stakeholders, initiatives, and elements within the urban system to enable them to work together effectively. They were the means of passing on information and so connected all the other functions of development organisation. This 'connectivity' contributed to the creation of a citywide network of urban food growers in three significant ways. First, by connecting community initiatives with resources and people which helped them to launch and develop, while existing initiatives shared information and experiences. Second, connecting initiatives with others via workshops and networking events encouraged collaboration and broadened outreach. Third, connecting initiatives to the public through multiple communication channels helped to increase participation. This social agency is underpinned by the citywide approach. It effectively saved time by establishing relations with innovation brokers, helping to identify skills gaps, support cross-sector relationships, and indorsed individual neighbourhood initiatives within a larger network to support their continuation.

It was shown how the Food Partnership's capability as a 'bridging organisation' between neighbourhood initiatives, local authority, and community groups was significant, especially for improving access to urban resources. This demonstrates the effectiveness of a lead coordinating urban development in partnership with frontline initiatives focussing on delivery. Creating a central point of contact needed to be underpinned by efficient communication and strong relationships between stakeholders. This informed opinions, built trust, and encouraged others to buy in to the programme. As such, relationships need to be developed with others who can solve issues that may arise, for instance finding out who is responsible for urban land. Developing cross-sector partnerships requires a concerted effort yet is shown to enable a more comprehensive action than individual projects or sectors might achieve on their own. While other cities face unique contextual challenges, my observations are proposed to have general application, expressly the combination of a systems-based approach with centralised coordination of urban development to connect people and spaces.

The provision of opportunities across the city contributed to the development of personal and social capabilities in the ability of local people to manage and resolve social issues, as well as to innovate individually and collectively to enhance quality of life. This was evidenced by the development of transferable skills at the neighbourhood level through volunteering and training activities to increase education and learning about cultivation, reflected by longer-term outputs in continuation of urban food growing. The outcome is that more residents have been enabled to access, both physically and in terms of their abilities, the benefits of this practice, which in turn affected existing lifestyle patterns and social interactions. Building material capacity is also observable in direct outputs in creation of physical and social infrastructure, especially by bringing new land and spaces into urban food production, and the widespread provision of volunteering opportunities and community events. Consequently, this urban framework provides the necessary environment for building individual and community capabilities to create self-governing development, innovation and change.

Localised actions to embed food growing as a collective spatial practice within the urban system are shown to be necessary to develop 'cultural capacity' (i.e. so that a practice is held as familiar), especially to advance broader community awareness, engagement, and ownership. Emerging green spaces became important social meeting places to provide the context for wider cultural changes within the city, with urban food growing as a social agent that builds community assets at a neighbourhood level. However, to facilitate long-term social and urban change raises questions on how to support social innovation initiatives that seek to scale-up as well as those that wish to self-sustain as community-driven. There is the issue of how to continue to promote opportunities and how to hand over activities to neighbourhood delivery partners. A solution to this challenge of providing citywide coverage may be found in a 'hub-and-spoke' distribution network to connect neighbourhood initiatives through centralised coordination (e.g. by local authority or partner organisation).

On analysing the reflection of socio-spatial development practice, the case indicates the Food Partnership's competence in demonstrating and communicating Harvest's outputs, including how they were realised to gain authority as an expert in urban food growing. In this sense, social action must be recognised across sectors to demonstrate its value and obtain wider credibility in order to advance innovation. Altering the existing environment of the urban system supported further approaches to embedding food growing practices locally; especially, in lobbying and negotiating with the local authority and other landowners to support long-term change. This is evident in the successful adoption of planning and development policy to support food growing in Brighton & Hove (see Chap. 8 case 'The Keep Community Orchard', Brighton & Hove). It is also important at the strategic level that there is recognition of the attention that the programme has brought to the city and for this reputation to be capitalised upon when developing future policy, strategies, and bids for resources.

For the Harvest programme's influence to continue to be shared it is necessary to maintain the momentum that has been generated. There is an ongoing need to measure and assess how and where learning has been adopted both locally and beyond. In order to advance innovation, I suggest that future dissemination of urban development knowledge should be supported to avoid any unnecessary duplication of agency. Networking will enable shared learning between cities working on similar agendas, as shown by formulation of the UK Sustainable Food Cities Network.

7.6 Interpretation of Social Innovation Pattern

This case involved several geographical scales of social innovation agency with movement back and forth between scales. The visual narrative map portrays a neighbourhood initiative occurring as a sub-component within the 'Harvest Brighton & Hove' urban development programme (Figs. 7.11 and 7.12). Both projects occur simultaneously and building on the previous case (see Chap. 6 'The Bevy Edible Garden', Brighton & Hove), identification of needs and development of social relations comes before provision activity.

VISUAL NARRATIVE MAP LEGEND

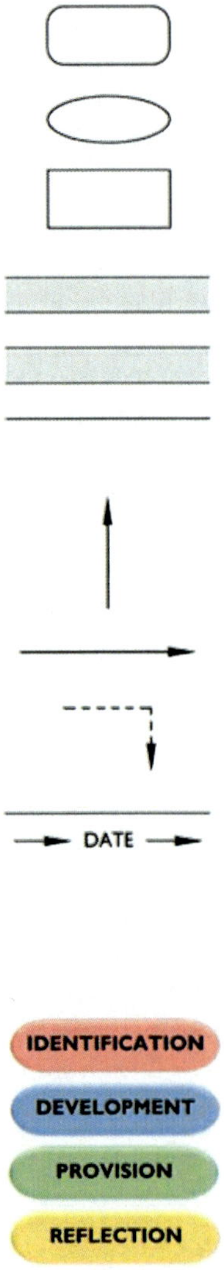

1. Polygons. The polygons on the visual narrative map describe the main elements of the case chronology relevant to social innovation.

- Rounded rectangles represent activities undertaken by the focal stakeholder
- Ovals represent events undertaken by the focal stakeholder
- Rectangles represent events and decisions outside the control of the focal stakeholder

2. Four horizontal stripes. Polygons are situated in relation to the contextual scale in which case activities and events are undertaken. Contextual scales are represented by the ascending horizontal stripes on the visual narrative map of: Neighbourhood, Urban System, and National System. The horizontal stripe to which all boxes are connected contains the base line activity of the core level of focal stakeholders that is indicated in bold.

3. Vertical arrows. Vertical arrows leading from one contextual scale to another indicate direct influences from events and activities on subsequent events and activities.

4. Horizontal arrows. Horizontal arrows within contextual scales indicate continuity of event or activity over time.

5. Dotted arrows. Indicate extended periods of time elapsing within the activity.

6. Time scales. The time scales along the bottom of each visual map vary in scale to minimise the size of maps whilst emphasising significant periods of activity and events.

7. Key social innovation processes. Key processes are represented by colour-coded polygons and situated next to the events and activities they relate to following case analysis according to the study protocols.

- Identification of human needs or societal challenges to sustainable development
- Development of social relations in systems or structures
- Provision of opportunity for social empowerment
- Reflection of socio-spatial development practice

The visual narrative maps can be read as an overview of the main events and activities within the case narrative and represent the complete case study incorporating narrative subplots. They provide a visual summary of the case study and through signifying key processes a graphical sequence from which sequence analysis can be undertaken.

(*Source* Author)

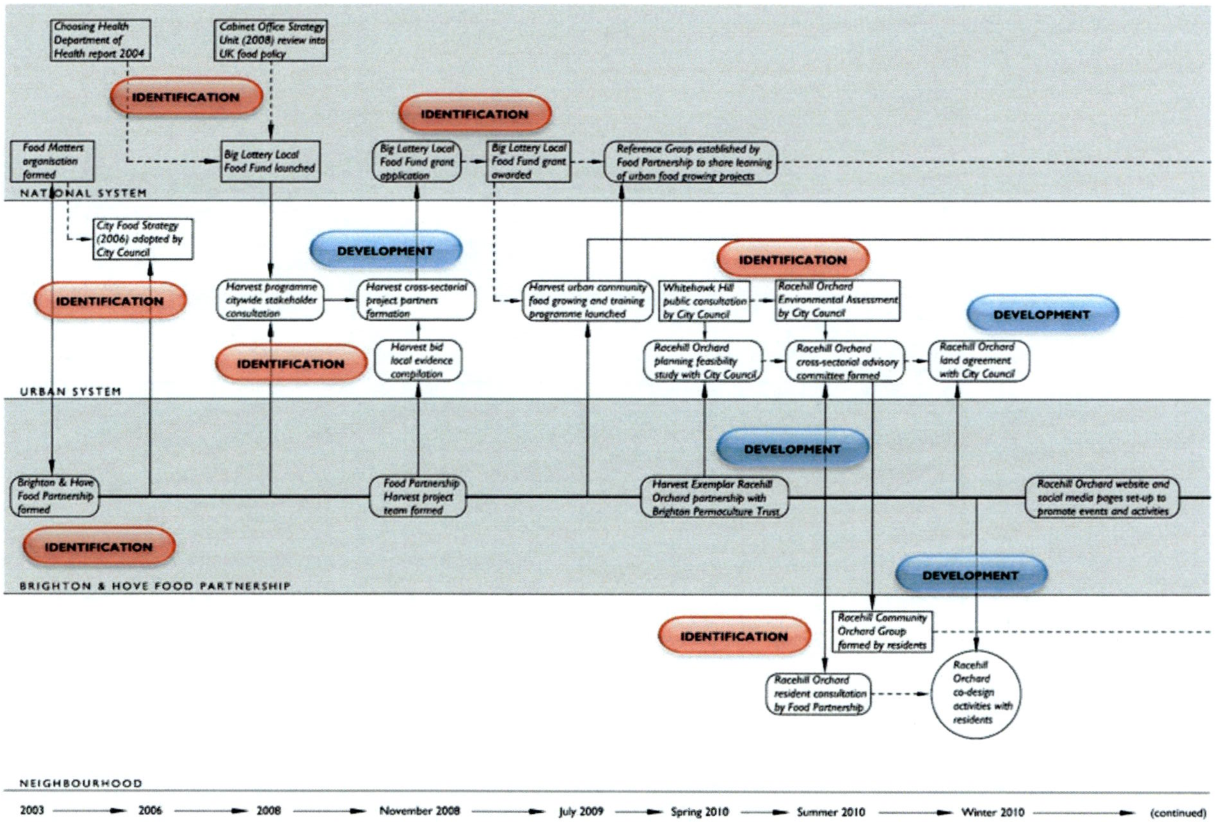

Fig. 7.11 Visual narrative map of the Racehill Orchard development with key social innovation processes indicated across contextual scales. Map sheet 1 of 2 (*Source* Author)

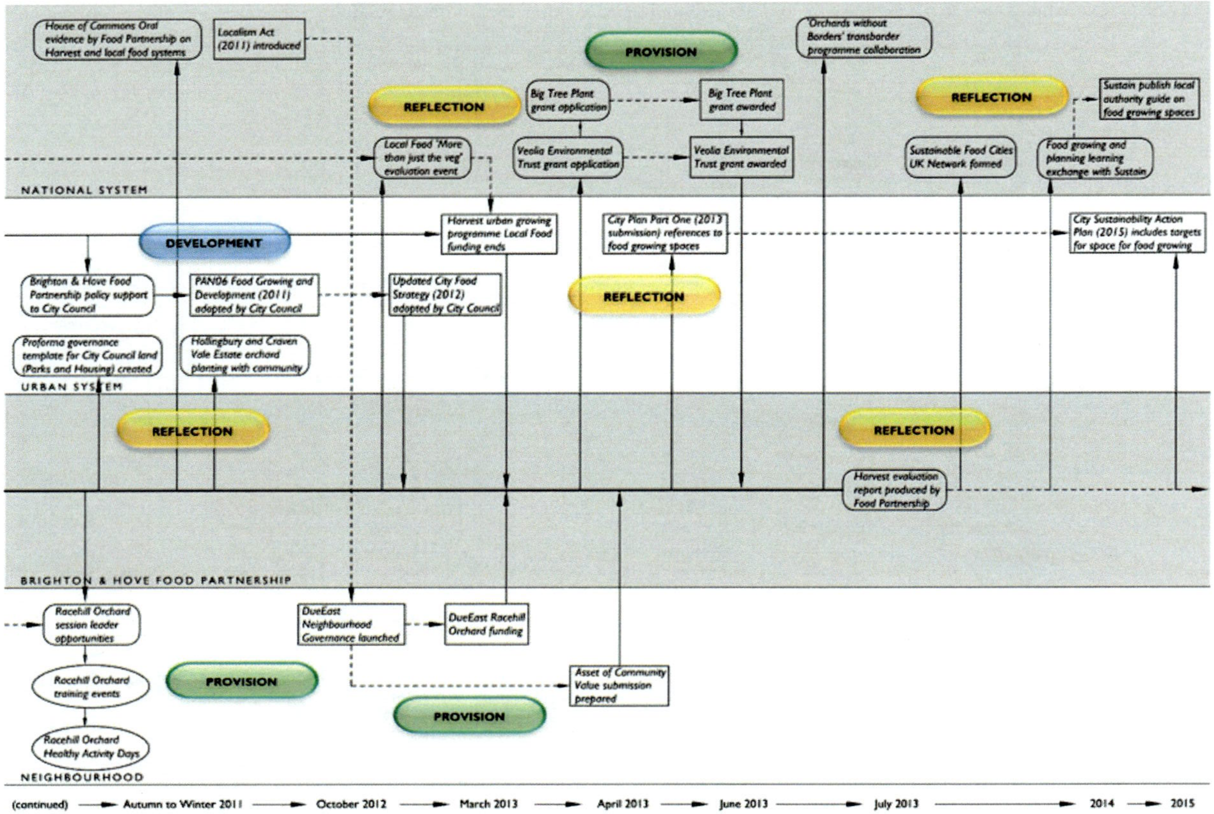

Fig. 7.12 Visual narrative map of the Racehill Orchard development with key social innovation processes indicated across contextual scales. Map sheet 2 of 2 (*Source* Author)

From Harvest's inception, identification processes can be observed at national to urban scales to make local food more accessible and increase community capacity. The development of relationships within the urban system follows in developing the necessary cross-sector delivery partnerships. Harvest's launch activates a sustained four-year period of provision activity across the urban system to develop community capacity. Comparable to the previous case (see Chap. 6), reflection activity again varies slightly from the proposed conceptual framework in Chap. 4 in terms of direction. This is because reflection is assumed before and after the conclusion of provision activity, evident in the national sharing of learning, and local influence through additional orchard developments. My observation does not affect the overall sequence proposed though demonstrates how this sharing of knowledge creates opportunities locally and in other cities through distribution of social innovation learning. Further reflection activity is evident at the end of the innovation sequence through meeting with partners to support policy lobbying, and externally via the national evaluation to Local Food and the formation of the Sustainable Food Cities Network to match the conceptual framework.

On closer examination this case exhibits the greatest reflection activity of all the cases due to the built-in strategy of 'mass-dissemination' (e.g. as a beacon project to inspire other cities and therefore start new cycles of innovation elsewhere). Significantly this reflection activity affects the urban system in terms of not just changing the city's material fabric, but also its socio-political environment to create opportunities for further identification and development activity, and thus subsequent innovations. In this sense, my finding helps to substantiate the claims of BEPA (2011) that socially innovative development can enhance society's capacity to act.

Table 7.1 provides a summary of contextual influences across spatial scales on development.

Table 7.1 Summation of contextual influences on Racehill Orchard development

Contextual scales	Influences on project development
National system (Macro scale)	• UK Big Lottery Local Food Fund beacon grant, set capacity improvement targets • UK Sustainable Food Cities Network creation, distribution of learning to other cities • Sustain, learning exchange • Food Matters advocacy sister organisation, political lobbying support and coordination between national and local strategies • Localism Act (2011), influenced experimentation with participatory processes • Veolia Environmental Trust grant, set guidelines • Big Tree Plant Government campaign grant, set guidelines • The People's Health Trust funding, set guidelines
Urban system (Meso scale)	• Brighton & Hove Food Strategy (2006, 2012), set local food growing targets • Sustainable Community Strategy (2006, 2010), set local food growing targets • Health inequalities in deprived wards, configured territorial aims • Harvest Brighton & Hove programme citywide stakeholder pre-consultation, influenced strategic brief • Cross-sector collaboration and service delivery partnerships, influenced agency • City Council provision of public land to develop for community food growing, catalysed urban development
Neighbourhood (Micro scale)	• Area of multiple deprivation, influenced need • Unexploited public land identified for development, created opportunity • Brighton & Hove Food Partnership resident consultation, influenced spatial and programmatic brief • Racehill Community Orchard advisory committee, influenced project, participatory framework between experts, intermediaries and residents • Interest in community food growing, Friends of the Racehill Community Orchard group formed

Source Author

References

BEPA (2011) Empowering people, driving change social innovation in the European Union. European Commission, Brussels

Brighton & Hove City Council (2011) Heads of terms for lease for Racehill Orchard, Whitehawk

Brighton & Hove City Council (2014) Brighton & Hove City snapshot. Brighton & Hove City Council, Brighton & Hove

Brighton & Hove City Council (2016) Community orchards growing in number. https://www.brighton-hove.gov.uk/content/press-release/community-orchards-growing-number. Accessed 29 Aug 2018

Brighton & Hove Food Partnership (2006) Spade to spoon: making the connections. Brighton & Hove Food Partnership, Brighton & Hove

Brighton & Hove Food Partnership (2008) Harvest Brighton & Hove workshop poster

Brighton & Hove Food Partnership (2012) Spade to spoon: digging deeper. Brighton & Hove Food Partnership, Brighton & Hove

Brighton & Hove Food Partnership (2014) Harvest Brighton & Hove evaluation summary 2009–13. Brighton & Hove Food Partnership, Brighton & Hove

Brighton & Hove Food Partnership (2015a) Racehill Community Orchard Report August 2015. Brighton & Hove Food Partnership, Brighton & Hove

Brighton & Hove Food Partnership (2015b) Brighton & Hove sustainable food cities award 2015. Brighton & Hove Food Partnership, Brighton & Hove

Brighton and Hove City Primary Care Trust (2006) Commissioning for health. The annual report of the Director of Public Health 2006. Brighton and Hove City Primary Care Trust, Brighton & Hove

Brighton Permaculture Trust (2013a) Orchard funding news. https://brightonpermaculture.org.uk/orchards/orchardnews/127-fundingnewsart. Accessed 2 Nov 2016

Brighton Permaculture Trust (2013b) Welcome to Racehill Community Orchard. Brighton Permaculture Trust, Brighton & Hove

Brighton Permaculture Trust (2015) About Racehill. https://brightonpermaculture.org.uk/orchards/racehill/aboutracehill. Accessed 28 Feb 2018

Brighton Permaculture Trust (2021) North Brighton Orchards planting. https://brightonpermaculture.org.uk/orchards-and-fruit/north-brighton-orchards-planting-of-70-trees-is-completed/. Accessed 6 Mar 2021

Cabinet Office Strategy Unit (2008) Food matters towards a strategy for the 21st century. Cabinet Office, London

Department for Communities and Local Government (2006) Allotments in England—report of survey 2006. Department for Communities and Local Government, London

Department for Communities and Local Government (2007) Indices of multiple deprivation 2007. Department for Communities and Local Government, London

Department for Communities and Local Government (2015) Indices of deprivation 2015: index of multiple deprivation and domain indices. Department for Communities and Local Government, London

Department of Health (2004) Choosing health making healthy choices easier. HM Government, London

Dimbleby H (2021) The national food strategy: the plan. DEFRA, London

Durrant R (2015) How can local initiatives help accelerate progress to sustainability in a socially inclusive manner? SPRU—Science Policy Research, Brighton & Hove

East Brighton Trust (2016) Community events at Racehill orchard. http://www.eastbrightontrust.org.uk/blog/read_165779/community-events-at-racehill-orchard.html. Accessed 31 Oct 2017

Else V (2008) Harvest Brighton and Hove business plan for the big lottery Local Food fund. Brighton & Hove Food Partnership, Brighton & Hove

Environmental Audit Committee (2012) House of commons sustainable food eleventh report of session 2010–12. House of Commons, London

Fainstein SS (2005) Planning theory and the city. J Plan Educ Res 25(2):121–130

Fainstein SS (2010) The just city. Cornell University Press, New York

Food Matters (2011) Brighton & Hove CSA: feasibility study interim report. Food Matters, Brighton & Hove

Health G (2015) Brighton & Hove Food Partnership: Harvest Brighton & Hove. Sustain, London

Harvey D (2009) Social justice and the city: revised edition. University of Georgia Press, Athens

Harvey D (2012) Rebel cities: from the right to the city to the urban revolution. Verso Books, London

Hayhow D, Eaton M, Stanbury A et al (2019) The state of nature 2019. The State of Nature Partnership, Sandy (Bedfordshire)

Iliffe F (2009) Sustainability cabinet committee agenda item 53: review of City Food Strategy 8 May 2009. Brighton & Hove City Council, Brighton & Hove

Kirwan J, Ilbery B, Maye D, Carey J (2012) More than just the veg growing community capacity through local food projects. CCRI, Cheltenham

Lefebvre H (1991) The production of space. Blackwell Publishing, Oxford

Moulaert F, Martinelli F, Swyngedouw E, Gonzalez S (2005) Towards alternative model(s) of local innovation. Urban Stud 42(11):1969–1990

OCSI (2007) Developing appropriate strategies for reducing inequality in Brighton and Hove. OCSI, Oxford

Soja E (2010) Seeking spatial justice. University of Minnesota Press, Minneapolis

Starr-Keddle H (2017) Community food projects and land in Brighton & Hove. Brighton & Hove Food Partnership, Brighton & Hove

Starr-Keddle H, Borrill V (2011) Evaluation of Racehill Orchard Consultation. Brighton & Hove Food Partnership, Brighton & Hove

Sustain (2014) Planning sustainable cities for community food growing. Sustain, London

Sustain (2018a) Food growing in parks: a guide for councils. Sustain, London

Sustain (2018b) How can the government's new national food strategy deliver on our shared aims ? In: Sustain. https://www.sustainweb.org/news/nov18_sustain_annual_gathering/. Accessed 4 Mar 2019

The Wildlife Trusts (2016) Local food—feeding the growth of local communities. https://www.wildlifetrusts.org/localfood. Accessed 20 Jun 2016

White R, Devereux C, Borrill V, Crocker J (2013) A growing community harvest evaluation report 2009–2013. Brighton & Hove Food Partnership, Brighton & Hove

Williams V (2019) Brighton and Hove Food Partnership: case study. https://www.foodmatters.org/case_studies/brighton-hove-food-partnership/. Accessed 7 Jul 2019

Chapter 8
Case Study of Social Innovation in Urban Food Spaces: 'The Keep Community Orchard', Brighton & Hove City Council, Brighton & Hove

Abstract In supporting sustainable development, Brighton & Hove City Council has adopted innovative policy which has led to this city experiencing an increase in planning proposals incorporating food growing. The Keep Community Orchard represented a significant influence on this policy-based innovation, specifically the PANO6 Food Growing and Development Planning Advice Note (2011), the first of its kind in the UK. A community orchard was integrated within 'The Keep' masterplan, a new public archive and historical resource centre for East Sussex (constructed 2010 to 2013). This civic development transformed a previously inaccessible site into a public green space featuring a productive area with permission to grow up to 0.53 ha. Contextually, in 2015 the North East Moulsecoomb estate bordering the site was one of the most deprived areas in the city and the 1,080th most deprived area in England out of 32,844, placing it at the national bottom 3%. This case study demonstrates how a social innovation intervention at the neighbourhood level is able to be translated and scaled-up into an urban policy, enabling the growth and replication of initiatives to effect far-reaching urban changes.

Keywords Urban food growing · Food resilience · Policy innovation · Planning policy · Sustainable development framework · One planet city · Social innovation · Brighton & Hove

The Keep Community Orchard is studied here as an innovative model of sustainable development integrating urban food growing and planning policy. As a local authority initiative, it is characterised as an output of policy-based social innovation (see Chap. 2). Development was coordinated by Planning Services and the Sustainability Team led by Francesca Iliffe, Brighton & Hove City Council Sustainability Officer. Iliffe advises on sustainability delivery while holding a remit for food planning in the city, particularly in supporting community groups to access land for cultivation.

Chronologically, this case study builds on the case 'Racehill Community Orchard (Harvest)', Brighton & Hove (Chap. 7). Several city stakeholders also featured in other Brighton & Hove cases, including: Francesca Iliffe and Mita Patel, Brighton & Hove City Council; Brighton & Hove Food Partnership; Bryn Thomas, Brighton

N. Ardill, *Growing Food in Cities*, Cities and Nature,
https://doi.org/10.1007/978-3-030-98475-5_8

Permaculture Trust; Warren Carter, Moulsecoomb Forest Garden and Wildlife Project; and Father John Wall of St Andrews Church, Moulsecoomb.

The following examines The Keep Community Orchard (henceforth The Keep Orchard) development from 2009 and the associated PANO6 Food Growing and Development Planning Advice Note (2011). Urban green infrastructure was instigated from 2010 onwards with a planning consultation process that led to provision of public land for community purpose.

8.1 Identification of Human Needs or Societal Challenges to Sustainable Development

Two principal identification activities are observed in this case. First, there was an opportunity to demonstrate the use of planning mechanisms through a prototype development to develop green infrastructure. Second, building on the first activity there was a need to generate policy to increase the amount of food grown within the city, create opportunities for food growing, and promote sustainable development (see Chap. 5 for a comparison of strategic policy approaches in the European city cases of Fredericia and Reykjavik).

An archetype for integrating food growing, planning and development

Brighton & Hove City Council (henceforth City Council) identified the need for a new county archive and a feasibility study was undertaken in 2007 (Allard and McSweeney 2010). Woollards Field on the urban fringe was selected for development. The site borders the Moulsecoomb estate (see Fig. 8.1), and historically was a fruit tree nursery, adjoining several market gardens and orchards before the housing estate development from 1920s onwards (Ordnance Survey 1937). Woollards Field was overgrown and of limited value prior to construction of The Keep (Fig. 8.2). During its planning feasibility, Francesca Iliffe, City Council Sustainability Officer recognised an opportunity to introduce productive spaces within development:

> It had a big open space on the site towards the Moulsecoomb end. I suggested when it came through planning pre-application discussions about 2006/7: you got lots of space there, why not put some fruit trees there? They said: no. Fruit trees? Why? There was no precedent for it at that time and I had no leverage to do that (Int_FI).

Although insupportable at the time, the proposal was later revisited amid greater City Council interest to integrating urban food spaces and sustainable development stimulated by local political changes.

In 2009 a pre-application planning consultation took place, and several recommendations were made. This resulted in a key obligation under Sect. 106 of the Town and Country Planning Act (1990) to make the proposal acceptable in planning terms and was based on a consideration of the Local Plan (2005) Policy QD20 Urban open space:

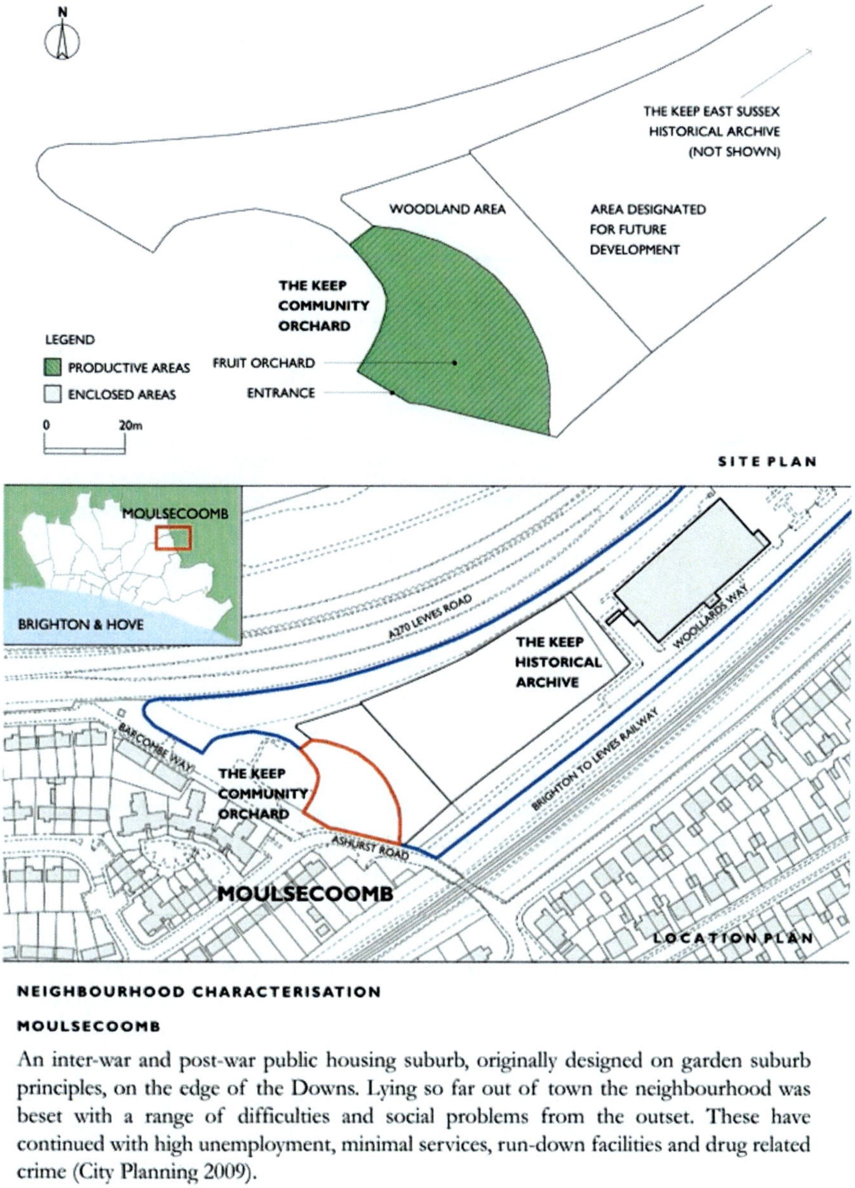

Fig. 8.1 The keep orchard site and location plan (*Source* Author)

In the 2015 Index of Multiple Deprivation (IMD) the Brighton & Hove 002D Lower Layer Super Output Area (LSOA) where the site is located was ranked 1,080 out of 32,844 in England, where 1 was the most deprived and 32,844 the least (Department for Communities and Local Government 2015).

Fig. 8.2 The keep public archive building (as built) (*Source* Author)

> Planning permission will not be granted for proposals that would result in the loss of areas of public or private open space that are important to people because of their recreational, community, historical, conservation, economic, wildlife, social or amenity value (Brighton & Hove City Council, 2005, p. 95).

The requirement to deliver public open space meant that an area of 0.53 ha of land was set aside for community purpose (Int_FI). Though Woollards Field had not been publicly accessible for several years, based on a predicted population rise and councillor advice, an area of public space was added to the masterplan as infrastructure for Moulsecoomb (Fig. 8.3) (Allard & McSweeney 2010).

Planning phases coincided with the introduction of the Archives for 21st Century (2009) Government policy. Public archives were intended to be a community resource and innovative model for education whilst promoting 'active participation in cultural and learning partnerships promoting a sense of identity and place within the community' (HM Government 2009, p. 5). Opportunities therefore needed to be provided to residents to participate and benefit in civic development, contributing to a later positioning of the community orchard as an educational community resource within planning (Int_FI).

A precedent development constructed nearby was also influential to the strategy of integrating cultivation spaces. The Brighton Aldridge Community Academy (2011) featured an outdoor classroom vegetable growing area and fruit orchard, though it also revealed issues around the need for ongoing management of spaces as the young trees were not being cared for. Consequently, a local group, the Moulsecoomb Forest Garden later took over tree management (Int_FI).

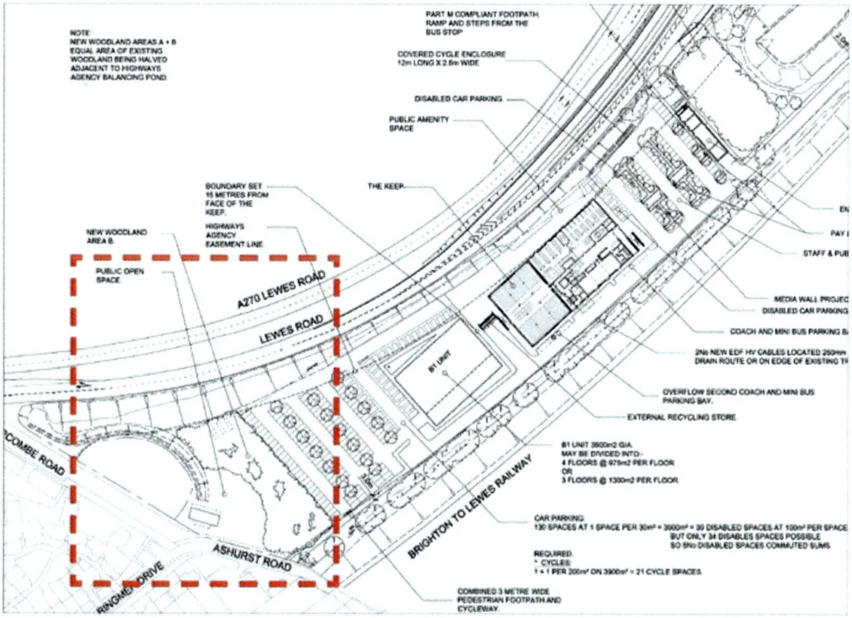

Fig. 8.3 The keep development masterplan (*Source* Morriss 2010)
Note 'Public open space' area as proposed shown boxed. The keep archive building is central to the masterplan

Policy-based innovation and sustainable development

From 2010, Brighton & Hove has been represented by Caroline Lucas, Green Party MP and in 2011 became the first city in the UK to elect the Green Party into local authority power. Subsequently, tackling inequality and working to create a fairer and more sustainable city became key priorities within the Corporate Plan (2011). The Plan outlined how the city would 'deliver the vision of the city's Sustainable Community Strategy (2006), by improving council services and through closer working with partners and residents' (Brighton & Hove City Council 2011a, p. 2). This increased top-down attention on sustainable development created an environment for changes in practice and policy (Int_FI). Sustainability became integrated within Planning Services as Mita Patel, Sustainability Programme Officer explains:

> [Because] it was the whole idea to embed sustainability into planning officers. [Iliffe] has run training sessions, so she is not the only expert on sustainability. When planning applications are reviewed, planners know what they are looking for in terms of sustainability—embedding that into the culture of the organisation (Int_MP).

Local policy built on the UK Strategy for Sustainable Development (1999); a catalyst for change in construction in setting out the Government's priorities for achieving sustainable development, including meeting needs in ways which deliver social progress that recognises everyone's needs (DETR 1999). Another policy driver was The Climate Change Act (2008) target for the reduction of UK carbon emissions by 80% by 2050 (HM Government 2008). In order to become a low-carbon economy, the country's food system was highlighted as a significant area due to 'huge environmental impacts [around 18% of UK greenhouse gas emissions are related to food production and consumption]' (Cabinet Office Strategy Unit 2008, p. i).

The City Council diagnosed local needs and substantiated challenges to satisfying needs in order to capitalise on a national agenda. Several policies of the concluding Local Plan (2005) related to green infrastructure and open spaces. However, there was no specific policy reference to the production of accessible 'community food growing spaces' aside from a vision for a 'quality local environment for all' and 'a relationship with the countryside [e.g. through local food]' (Brighton & Hove City Council 2005, p. 16). This oversight was interpreted as a need for specific policy to support community food growing and innovative approaches to sustainable development (Int_FI).

The Keep's development was concurrent with the 'Harvest Brighton & Hove' programme (2009–2013) led by the Brighton & Hove Food Partnership (henceforth the Food Partnership) which shared a common aim to develop policy and explore how planning could be utilised to support urban food growing (see Chap. 7 case study 'Racehill Community Orchard', Brighton & Hove). The integration of food growing spaces within development presented a planning archetype and 'a really good model for other developments to adopt. It was quite an easy thing to replicate, while at the same time it was a big attraction' (Int_FI). In lobbying for policy change, The Keep was positioned as 'a case of good practice' in supporting the generation and adoption of the PANO6 Food Growing and Development Planning Advice Note (2011), and updated in 2020 (Brighton and Hove City Council 2020).

Brighton & Hove then became the first local authority in the UK to adopt planning guidance encouraging applicants to integrate food growing within developments (Brighton & Hove City Council 2011b). The Planning Advice Note (2011), developed by Iliffe, the Food Partnership and Food Matters, was intended as model guidance that might also be adopted in other cities (Fig. 8.4). Iliffe says: 'It was to increase visibility; make people see that it's demystifying food growing so they saw it in various settings. People would have more access to food growing and more food growing in the city' (Int_FI). Furthermore, this guidance sought to build on the visibility and momentum of Harvest, presenting an opportunity for urban change:

> There was ongoing policy securing food growing. It was written to inspire both planning authorities and developers to be innovative and creative, but there are no actual requirements for anyone to do anything which was crucial in terms of getting it approved by the Council (Int_FI).

The document is presented as a Planning Advice Note (PAN) in contrast to a Supplementary Planning Document (SPD). While this means it holds less power (i.e. no 'requirements'), there was a simpler adoption process and technical guidance

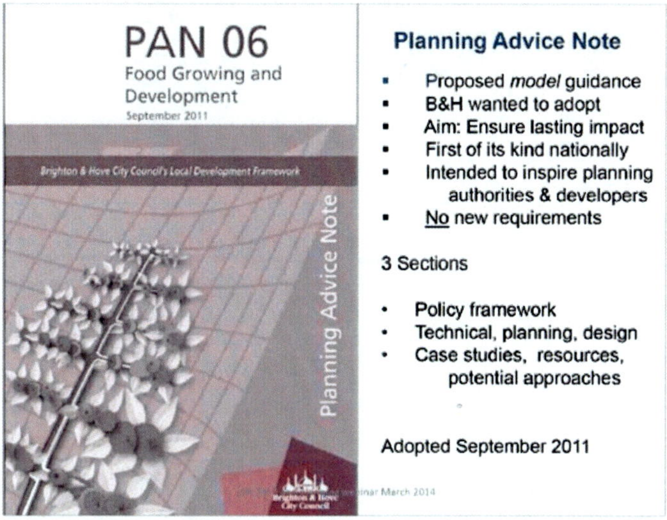

Fig. 8.4 Brighton & Hove PAN06 Food Growing and Development Planning Advice Note (2011) (*Source* Iliffe 2014)

could be introduced without satisfying statutory processes. Iliffe says: 'you could not really introduce the requirements unless it was in the development plan document [City Plan Part One]. We had not adopted our new development plan document until 2016—so we would have had to wait five years' (Int_FI). The strategic intent was to show planning applicants and developers that the city wanted more food growing spaces to support sustainable development.

A Planning Sustainability Checklist was also developed at the same time to improve city sustainability. The Checklist provides planners and developers with an overview of issues to consider in processing applications. This process meant:

> Everyone putting in a mixed-use planning or residential planning application had to fill in this online form. We introduced a page on food growing. So, we could do some monitoring and see whether there was any impact on adopting the PAN, but also to indicate to people that we wanted them to do this, to be delivering something (Int_FI).

The food growing section asks: 'Is there provision for food growing included on the development site?' and requests information regarding food growing areas and number of trees (Fig. 8.5). The checklist's utility however is restricted to gathering data on how policy is being implemented at the application stage. Significantly, it does not track or corroborate information concerning realisation of these productive spaces in the city.

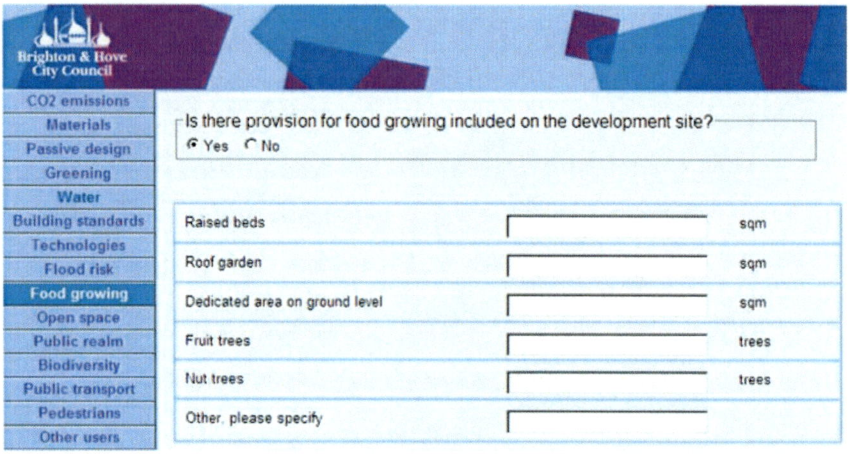

Fig. 8.5 Brighton & Hove planning sustainability checklist 'food growing' section (*Source* Iliffe 2014)

8.2 Development of Social Relations in Systems or Structures

Consultations were central to the development of relations in formulating policy through joint lobbying and a public consultation facilitated participatory planning. The influence of intermediaries in brokering relationships within the urban system is also evident in this case. Several local policies already included references to food growing owing to Food Partnership consultations and internal City Council support. Iliffe points out: '[Because] of the Food Partnership responding that we have that aim within the Sustainable Community Strategy (2006). It goes into the policy document and then it comes a hook for us to do other things' (Int_FI). Commenting further, Patel suggests:

> [This relationship] is a two-way thing. From the community sector perspective, the Food Partnership are very involved and influential. [Iliffe] is pivotal and has a good role in being able to push that agenda, having constant dialogue with Food Partnership, always looking out for new projects to develop […] influence facilitating those discussions between the community sector organisations, individuals and councillors (Int_MP).

These collaborative relations were significant to applying pressure, both from stakeholders below and above (e.g. communities and local institutions), for planning and development policy at the strategic level to increase urban food growing opportunities.

Public consultation and approaches to participatory planning

In early 2010, a public consultation of The Keep planning application was held. This involved three core elements of 'interaction' as proposed by OECD (2006). Actions included: (1) Notification to publicise the matter to be consulted on; (2) Consultation via a two-way flow of information and opinion exchange; and (3) Participation,

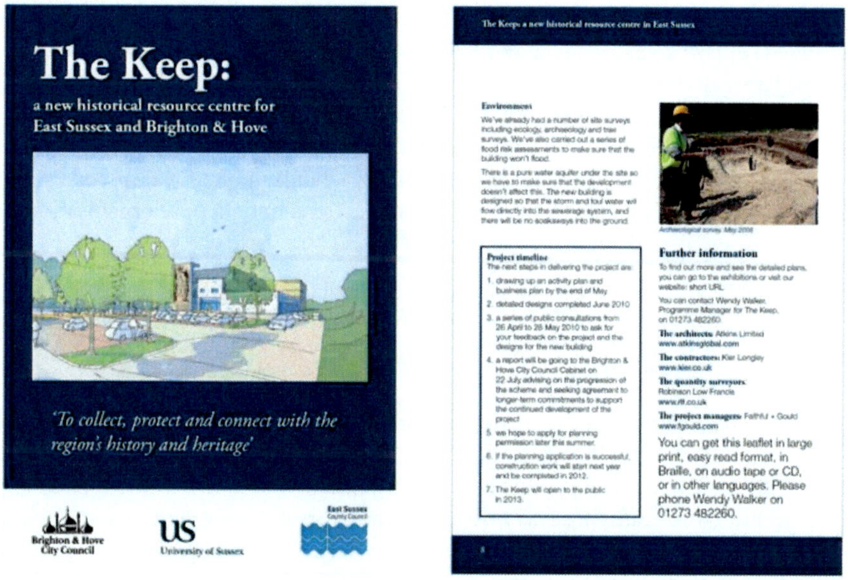

Fig. 8.6 The keep development public consultation newsletter (*Source* Morriss 2010)

involving interest groups in the drafting of strategies or policy. A series of public exhibitions, newsletters, and questionnaires were undertaken (Fig. 8.6), while discussions with local councillors provided insight and advice on neighbourhood consultations (Kier Regional 2010). The councillors prepared a list of representatives to participate in consultations and identified several areas of concern for the community, including loss of Woollards Field as a potential amenity space. A resident focus group was then held in March 2010 at St George's Hall, Moulsecoomb (Morriss 2010), while further meetings were held with the Moulsecoomb Local Action Team from April 2010 (Kier Regional 2010).

Through consulting those directly affected by development, the participatory approach to planning increased levels of transparency and thus contributed to 'socially just' decision-making processes in urban spaces as argued for by Fainstein (2010) and Marcuse (2009). Father John Wall, Vicar of St Andrews Church, Moulsecoomb was one of the local participants. As previously introduced (see Chap. 6 case 'The Bevy Edible Garden', Brighton & Hove), Father Wall was a prominent member of the local community and his 'Leaders Lunches' hosted at the vicarage supported local networking and relationship building. Father Wall describes how:

> The County Archivist used to come along to my lunches and did a presentation. We had a tour around The Keep. I went two or three times while it was being built. I was involved for two reasons: (A) Because I was the vicar of Moulsecoomb, and it was happening on my patch; and (B) Because I was the mayor's chaplain (Int_FJW).

Father Wall's intermediation between stakeholders and across organisational levels to support social innovation is apparent in creating opportunities and relationships. Through developing grassroots relations with Moulsecoomb Forest Garden he connected to Brighton Permaculture Trust who made it known '[that] we are trying to find public spaces to plant apple trees, we have got funding for it. So, I thought where better to have an archive of Sussex apples than actually in the East Sussex County Archive' (Int_FJW). Father Wall used his connections to link these organisations to the City Council, and suggests:

> I brought the Brighton Permaculture Trust, who had the tree money, together with the people who were running The Keep. I was saying: this is public land, it is to do with an archive, you have got the funding, the knowledge, and you want to plant Sussex trees […] They got together and liked the idea and it just happened (Int_FJW).

Father Wall's idea had been partly inspired by an elderly resident who informed him that Woollards Field was a site of 'scrumping' after World War II; providing 'free fruit' during an era of food shortages when rationing still commonplace. This local pastime of scrumping was utilised by the project team as a heritage link and cultural brief for the public open space within the civic development's masterplan, being reconfigured during planning as a community orchard (Int_FI).

A state-civil society partnership

The contribution of civil society groups aligned with state steering to coordinate processes of social innovation created the local conditions for hybrid stakeholder partnerships to emerge. A formal agreement for the orchard management was drawn-up and included within planning documentation. It meant the community orchard would be co-managed by local organisations Brighton Permaculture Trust, as project lead, and Moulsecoomb Forest Garden as neighbourhood-based partner (see Chap. 7 case 'Racehill Community Orchard' and Chap. 6 case 'The Bevy Edible Garden', Brighton & Hove). Activities were to be overseen by City Parks (City Council) in collaboration with the local organisations (Int_BT). From spring 2012, several discussions transpired between the City Council, Iliffe, and project partners to coordinate the orchard's development as Thomas explains:

> Site meetings with contractors and all sorts of people which is quite unusual […] it is not a common agreement. It is the only one where we have actually had to sign up to say we are going to care officially […] because it was part of a planning document it all got written down in far more detail than anything else we have been involved with (Int_BT).

In terms of cross-sector working relationships, Warren Carter, Moulsecoomb Forest Garden points out that having support within the City Council was important, especially as Iliffe 'was really good with The Keep when we did the orchard. That happened so smoothly I can't quite believe it' (Int_WC). Bryn Thomas, Brighton Permaculture Trust adds: 'a very good hard-working officer […] that choose to push stuff through' (Int_BT). This relationship reinforced an emerging collation with Iliffe also bringing on board the Food Partnership to support and promote the project. Vic Borril, Director of the Food Partnership explains how:

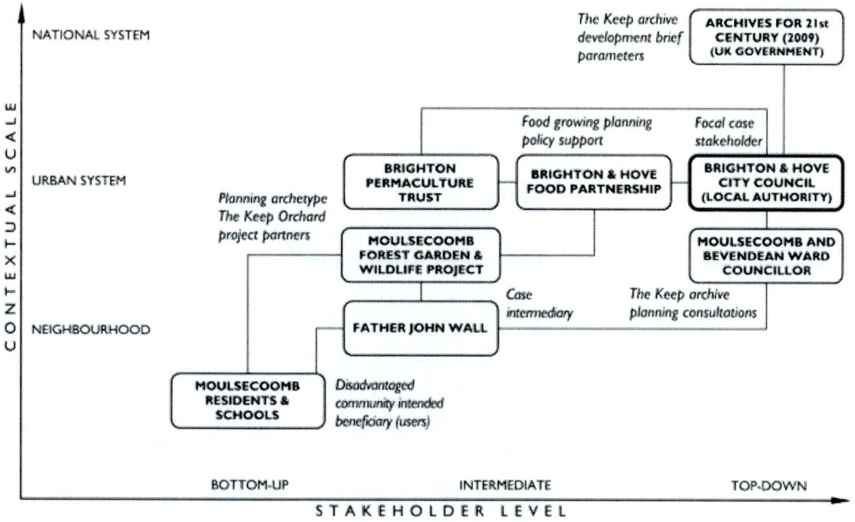

Fig. 8.7 Stakeholders and their relationships in The Keep Orchard development (*Source* Author) *Note* Focal case stakeholder outlined in bold

> Local people asked the Harvest partners if the landscape area around The Keep could be used as a productive landscape. Community food projects and the Council's planning team and City Parks have worked together to make this happen and this is a great example of the City Council's planning policies to support food growing in action (Brighton Permaculture Trust 2012).

In short, the development exemplified collaboration and stakeholder partnerships within the urban system through integration of community food growing spaces, planning policy and development.

Figure 8.7 highlights key stakeholders and relationships in The Keep Orchard.

8.3 Provision of Opportunity for Social Empowerment

The delivery of a neighbourhood resource offering educational opportunities to disadvantaged residents to learn about food growing, environmental issues and sustainability in the community is evident in this case. Furthermore, through the formulation and adoption of policy supporting cultivation spaces this increased the distribution of neighbourhood opportunities to the urban scale.

Increasing opportunities within the neighbourhood

The Keep Orchard was planted from late 2012 to early 2013 by pupils from Moulsecoomb Primary School and Brighton Aldridge Community Academy (Fig. 8.8). Historic research was undertaken to ensure heritage tree varieties were planted to aid their conservation and an apple archive was created. The elderly resident who

Fig. 8.8 The Keep Orchard's planting with local schoolchildren (*Source* Iliffe 2014)

spoke with Father Wall prior to development 'had written a story about scrumping, about a local fruit tree variety called White Transparent. So, we planted that variety at the orchard and that letter is now held at The Keep as well' (Int_FI). In this regard, the production of a community orchard as a cultural object supports the positioning of residents within the environment around a specific location, providing orientation to support identities. Planting activities were undertaken here by the community in contrast to other landscaping to foster territorial ownership. Furthermore, Peter West, Moulsecoomb Councillor and City Council Environment Committee Chair, proclaimed the orchard's social and cultural value as:

> So many of our native orchards have disappeared, so it is wonderful that we are able to put one back. It also echoes the purpose and ethos of The Keep which will bring local archives together in one place for people to discover the history and heritage of Brighton & Hove and surrounding area (Brighton Permaculture Trust 2012).

The Keep Orchard was devised as a community space to deliver recurring public events and activities and opportunities to improve access to urban resources. From 2012 to 2015, services were delivered by the Moulsecoomb Forest Garden and featured awareness-raising with children who might otherwise be totally disconnected from how and where food is produced. For example, students from Brighton Aldridge Community Academy participated in outdoor learning activities. At the time, Natasha Silsby, Community Manager, Brighton Aldridge Community Academy stressed how productive spaces such as these 'present such rich learning opportunities and health benefits for our students and the local community' (Carter 2012). Vanessa Brown, Children and Young People Cabinet Member, the City Council added how they are 'an example of sustainability in action and will enable future generations to make the connection between spade and spoon' (Jordan 2012, p. 3). Horticultural training sessions were also held for residents and their participation was encouraged in the ongoing care of the community orchard. Public events, including 'Apple Days' were planned to coincide with the national event held each September around harvest festivals. This event uses the apple as a symbol of physical, cultural and genetic diversity of a locality to demonstrate to the public that it is possible to influence place-based change (Brighton Permaculture Trust 2018).

Despite the initial enthusiasm for the community space and activities, from 2016 onwards The Keep Orchard ceased operations. It was no longer accessible as a public space and became overgrown, threatening its long-term continuation (Fig. 8.9). Although reasons are undecided, the City Council are held responsible in-part through installing a barrier to the site 'to keep travellers off, fly-tippers, or both', especially as the site was vulnerable to misuse as an open space situated on the urban fringe (Int_BT). Locally, this situation is prevalent with residents having expressed concerns about travellers in relation to green spaces on the city's edge (Brighton & Hove Food Partnership 2015). It is also suggested that the City Council failed to undertake their management responsibilities. Carter says:

> It is overgrown because the Council has not done what it is supposed to do in terms of its maintenance [...] I hassle the Council to mow the grass and clear the rubbish. I had hoped

Fig. 8.9 The Keep Orchard site, late 2016 (*Source* Author)
Note The Keep archive building is seen in the distance. Orchard apple trees, overgrown wildflower meadow, and site entrance with installed barrier are also shown

> The Keep being more active in that […] no one wants to take ownership and look after it (Int_WC).

Thomas adds:

> [The orchard] was one of the ones that was in some ways the idea of the great and the good and slightly suffered from lots of people who were thinking it was a good idea but no one on the ground saying: yes, I want to drive this forward (Int_BT).

As such, the initiative has reinforced the need to establish an ongoing duty of care over public green spaces, especially when involving hybrid governance structures, to ensure individual responsibilities are defined and stakeholders held accountable for what was agreed upon.

Implementation of citywide planning policy

Assessing the development of The Keep Orchard, the provision of social value within the neighbourhood appears limited in sustaining access to public space and thus active participation of residents. Yet at the city level, it is suggested that this development as an archetype for planning policy supported the wider provision of opportunities to participate in urban food growing whilst increasing access to material resources. The archetype's influence is evident in the Planning Advice Note (2011), and city-wide developments integrating cultivation spaces. Several proposals included food growing or made Sect. 106 contributions to the development of offsite spaces (see Figs. 8.10 and 8.11). For example, the Brighton & Hove Circus Street mixed-use

Fig. 8.10 Circus Street, Brighton development and food growing proposals (*Source* Iliffe 2014)

Fig. 8.11 Development integrating food growing in Brighton & Hove following the Planning Advice Note (2011) (*Source* Author with data from Iliffe 2014)

scheme and innovation quarter (completed 2021) features a 'Spade to Spoon' growing hub with food integrated throughout the site. The development takes inspiration from the derelict municipal fruit and vegetable market in the city centre where it is located and aims to create a productive quarter featuring multiple food growing spaces. This inner city area has been traditionally underserved in green space and proposes two large, public squares where food growing will be visible, with Brighton Permaculture Trust providing the training services in the design, procurement, and aftercare of fruit trees. Learning from previous developments, the scheme incorporates a head gardener to oversee a management structure encouraging community participation, and will connect to local food networks, such as the Food Partnership to further opportunities and access to resources.

8.4 Reflection of Socio-spatial Development Practice

A prominent reflection activity observed in this case concerns stakeholders learning from experimentation with policy-based innovation, including the advancement of a sustainable development framework integrating community food growing as a core principle of local policy.

Brighton & Hove's PAN06 Food Growing and development (2011) Planning Advice Note was monitored to measure its impact. In the 12 months following the City Council's adoption, 44 out of 98 planning applications (a rise from 1 to 45% since adoption) included 712 m^2 of food growing space within development proposals (Iliffe 2014). Consequently, Iliffe proposes:

> There is something in the psyche of planning and construction mindsets in Brighton & Hove. It is accepted that there is going to be food growing, this is part of what is to be expected and to be delivered (Int_FI).

Thomas adds:

> [By] just putting the questions to developers: have you considered community food growing? Have you considered biodiversity improvements? It has helped to put it in developers' minds even though there is no requirement […] When they answer the question, they are much more likely now to answer with a yes (Int_BT).

The planning guidance was recognised in 'Innovation' awards, regionally and nationally (Brighton & Hove Food Partnership 2014). Discussing the role of food growing to enhancing health and wellbeing in Brighton & Hove, Phelim Mac Cafferty, Councillor and Chair of the City Council Planning Committee, proclaimed:

> We have been awarded 'Healthy City' status in recognition of our commitment to reduce health inequalities. We believe that locally grown food has a crucial role to play. Today in Brighton and Hove spaces for food growing in new developments are being secured because of our ground-breaking planning policies and guidance (Sustain 2014, p. 16).

Iliffe points out how the City Council has 'done a lot of work encouraging developers to green their sites. The guidance is part of trying to improve our overall green infrastructure' (Jordan 2012, p. 2). Jeremy Iles, Chief Executive Federation of City Farms & Community Gardens stresses Brighton & Hove should serve as an example to other localities because 'they have a great partnership between community and local authority, which I think has helped lead to the planning guidance. But there are only a few isolated examples of this kind of innovative thinking' (Jordan 2012, p. 2). As such, the legibility of this collaborative approach is argued to influence changes in meaning communicated by the city's reputation and physical form:

> We had such a lot of interest in Brighton and Hove in terms of the work being done with the food system [...] it has an impact on the image of the city. The Council has encouraged and supported any planning authority that comes to us and says: we really like what you have done with the Food Strategy, the Food Partnership, the Planning Advice Note, would you help us? [...] particularly on an agenda like this, sharing so that it can happen elsewhere (Int_FI).

The city's approach to developing food growing spaces and provision of support and know-how from one organisation or place to another has distributed knowledge. Owing to the attention Brighton & Hove has received this diffusion process is observable in the London councils of Kingston, Kensington and Chelsea, and the cities of Leicester, Bristol, and Belfast adopting similar planning policies which encourage community food growing and sustainable design to reduce carbon emissions and promote health (Sustain 2014). Learning from Brighton & Hove was further shared to other cities nationwide and internationally through the Food Partnership and the UK Sustainable Food Cities Network.

Challenges to social innovation in urban food spaces

Evaluating learning and reflection activities in this case reveals two challenges to continuing social innovation. First, in terms of planning schemes realising productive spaces:

> [While the City Council] had a lot of success in terms of what is being proposed, it shows that we don't know that these schemes are actually being built as this and will deliver these things for a while because there is a three or four-year time limit between something coming into planning, being built out and occupied (Int_FI).

The need for further monitoring and enhanced policy for enforcement is confirmed by examining development outcomes. The Food Partnership's Development Officer Starr-Keddle suggests 'there are issues with the planning advisory now; where people design spaces incorporating gardens and then they don't actually materialise' (Int_HSK). Likewise, Iliffe explains how the City Council:

> [Attempted] to monitor along, but it was a bit too early because the schemes may not have been built out [...] Food Partnership monitoring showed there was big gap between what was being proposed and what was being delivered. We know there is work to be done on that (Int_FI).

Furthermore, City Council monitoring was restricted:

> Because we usually monitor by desktop monitoring. We do not have the resources to go out on site and monitor anything but the headlines of developments on site. There is not an overarching framework (Int_FI).

Consequently, there is a suggested need to lobby the City Council to include additional food growing 'requirements' within the City Plan Part Two (forthcoming). Another strategy would be to experiment with a Supplementary Planning Document to increase the power of the existing Planning Advice Note (2011) from advice to a 'material consideration' for determining applications. Kingston and Leicester for example suggest a replicable model through the introduction of Supplementary Planning Documents that obligate developers to integrate food growing in the interest of sustainable development (Sustain 2014). Iliffe suggests 'it would be helpful to do more funding. You need to make these projects self-sustaining. Getting funding to seed them is one thing but you also need to have the interest to keep them going. So, that's the other challenge' (Int_FI).

Second, the management of productive spaces within developments is significant as 'you have to have people looking after those spaces that is the issue; we are still learning from that' (Int_HSK). Thomas says: 'one of the things about community orchards; everyone thinks about who is going to plant them, but what we normally look for ideally is a group of people who want to plant it but also a group of people who want to do the ongoing maintenance' (Int_BT). As such, development monitoring has revealed 'where there is an ongoing duty of care from the owners of the building, whether student residential hall, social housing, or residential care home you have a chance of getting food growing because of the ongoing responsibility of the building manager' (Int_FI). A process of learning and adapting behaviour is thus suggested within the case, specifically in the Brighton & Hove Circus Street development where there is now a new requirement for a head gardener to oversee community participation in land management.

Brighton & Hove sustainable development framework

The case helped to demonstrate the need for a sustainable development framework for Brighton & Hove, presenting an opportunity to expand green infrastructure. In 2013, following the adoption of the Planning Advice Note (2011) the city adopted the 'One Planet City' sustainable development framework (Brighton & Hove City Council 2015). The framework seeks a reduced ecological footprint and local stakeholders are stimulated to develop tailored solutions to urban issues, whilst innovation and experimentation are promoted, and stakeholder collaborations enabled through local partnerships (Bioregional 2017). Sustainable city frameworks such as One Planet are considered by Joss (2015) to be positioned to disrupt and replace conventional urban planning mechanisms by setting out the principles, establishing rules, and ultimately redefining governance approaches to innovation and sustainable development. For the City Council, the city's framework is 'a way of communicating sustainability as it provides a comprehensive framework of structuring sustainability through 10 core principles. It makes people think about sustainability broader than just about the environment; it is about the social elements, the health elements as well' (Int_MP).

The principles of the city's sustainable development framework connect to relevant City Plan Part One (2016) policies and are integrated as a broad set of actions within the Brighton & Hove Sustainability Action Plan (2015). The Plan coordinates the distribution and measurement of initiatives supporting sustainability being delivered by the City Council, partners, and individuals across the city. Furthermore, it emphasises using community assets in people and places to design public services collaboratively (Brighton & Hove City Council 2015). This approach triggered the local formation of the City Sustainability Partnership, a cross-sector initiative comprising various stakeholders taking on the role of principle leads for the delivery of various thematic areas. 'Local and sustainable food' is now a core principle and led by the City Council Planning Services with support from the Food Partnership (Brighton & Hove City Council 2015, p. 38). This collaborative arrangement connects to the idea of 'procedural justice' offered by Low & Iveson (2016) in the negotiation of decision-making processes shaping public urban spaces and their use. Furthermore, the Plan supports green infrastructure with policy and Iliffe stresses that 'anything that is not being achieved in the Food Strategy, or any areas that are blocked, we try to throw them into that document' (Int_FI). Significantly, it sets key actions to increase the number of food growing projects in the city, including: 'Deliver: 5 new food growing projects on council housing land by 2016; at least 2 community orchards of over 5 trees each year starting 2014; and one further demo garden every year to 2016' (Brighton & Hove City Council 2015, p. 44).

A cyclical process of community consultation and formulation of planning and development policy is also evident. Cross-sector stakeholders fed back learning into the City Plan Part One (2016) and City Plan Part Two (forthcoming). This process facilitated a policy forum 'to develop a joined-up framework for seeking solutions [...] In doing so it attempts to engage the whole community across a diverse range of sectors in a challenging and innovative way' (Brighton & Hove Food Partnership 2006, p. 3). Emma McDermott, Head of Communities, Equality & Third Sector explains here how the City Council consult with the community sector on policy adoption:

> They are undertaking quite an extensive consultation process for our City Plan Part Two. That happened during Part One. Now they are revisiting that; creating a new plan and many individuals in the city, particularly in the environmental sector have a strong interest. It demonstrates an appetite to include the voice of the community sector, making sure that this is representative of all the different sectors in the city (Int_EMD).

As a result of this engagement, several references 'to support and promote local food growing' were integrated within the City Plan Part One (2016), specifically in polices: SA6 Sustainable Neighbourhoods; CP8 Sustainable Buildings; CP16 Open Space; and CP18 Healthy City (Brighton & Hove City Council 2016). The application of policy is demonstrated by the Brighton & Hove Toad's Hole Valley development (Fig. 8.12), with the City Plan Part One (2016) stating that this major development must deliver an 'Ecology Park comprising: two ha of public open space; green infrastructure integrated throughout the site; and that 0.5 ha should be set aside for food growing by local residents within and in neighbourhoods near to the site' (Brighton & Hove City Council 2016, p. 87).

Fig. 8.12 Toad's Hole Valley, Brighton & Hove masterplan (*Source* Brighton & Hove City Council 2016) *Note* Ecology Park shown left

8.5 Case Discussion and Conclusions

In this section I discuss the interactions between key social innovation processes and summarise my findings. I conclude the following section with a reinterpretation of this case in key processes and spatial scales of analysis.

In summary, I analysed how a local authority sought to integrate food growing within local planning and development through policy-based social innovation. The City Council built on existing cross-sector partnerships whilst further developing stakeholder relations to formulate planning guidance that led to a growth in applications integrating spaces for food growing. This strategy created social value through the development of urban green infrastructure, increasing opportunities for community participation in food growing, and advanced a sustainable development framework supporting cultivation spaces in Brighton & Hove. Meetings took place with multiple stakeholders both within the city and nationally to disseminate policy knowledge.

The identification of needs was shown to be triggered by a change in city leadership and a shift towards environmental priorities, creating political opportunity. This 'window of opportunity' for strategic intervention and urban transformation existed due to the city experiencing an openness for change. The City Sustainability Officer, as agent of change recognised this specific event presented an opportune time to experiment with (and ultimately advance) innovation in planning and development. This opportunity for change is suggested to be situational because urban governance is susceptible to cyclical adjustment, affecting political agendas. In this sense, political structures are continually shifting in the interaction of local politicians and urban stakeholders, and therefore political opportunity was linked here to a specific context (e.g. the UK Green Party assuming local authority power in 2011).

Within the local environmental there existed a momentum for urban change amid shared interest and encouragement of food growing as cooperative relationships existed between local authority and local organisations through partnerships and associations. Identification activity was shown to attempt to activate change through existing political structure and processes. It was therefore necessary to foster political buy-in to seize political opportunities and activate this urban change. Here a shared understanding helped to legitimise and stimulate collective action at the strategic level. Moreover, there was a need to define the existing challenges clearly and persuasively; articulate why urban change was necessary; identify what changes were required and how to go about implementing them.

An analysis of how social relations were developed suggests that collective lobbying influenced local policy-making. Coalitions of interest groups were mobilised in pursuing the same policy objective: to increase local food growing practices and support sustainability. As such, the City Council offered decision-making influence for the provision of local insight and development support from the Food Partnership in behaving in a goal-oriented fashion. Practitioner organisations can therefore offer policy-makers a shortcut to local knowledge required to advance urban development processes because they are well equipped with expertise and information in their field (e.g. food systems, health and education). Furthermore, the City Council in this instance was receptive to this way of working because its time and resources were limited, and their agency restricted in gathering the necessary insight to develop effective innovation solutions to meet local needs.

Key to gaining institutional access to influence policy makers directly was face-to-face contact with and influence on representatives. Insider lobbying was aided by building cross-sector relationships and having a 'change agent' within the local authority with shared interest, furthering political support (e.g. Iliffe, City Council Sustainability Officer and Food Partnership Board Member). This strategy contrasts with more indirect, external approaches that aim to apply pressure on authorities by gaining public support through online campaigns, or by grassroots mobilisation and demonstrations. Therefore, the contextual setting and governing environment are suggested here as being important in the understanding of how advocacy can be integrated into the political process to influence local policy. In this case, lobbying activity was conditional on the local authority setting and use of shared decision making approaches, e.g. policy consultations undertaken between local authority and non-governmental organisations. The creation of participative governance structures helped in enabling various urban agendas and goals to converge so that they were not incompatible. Moreover, these participative structures enabled multiple interest groups to provide inputs to policy-making and to cooperatively shape urban development.

The provision of opportunities in this case was shown to be advanced by building on existing experience in order to scale-up an innovative neighbourhood solution to an urban policy. The individual, experimental initiative informed local authority learning and change at the institutional level for wider-reaching impact. The Keep Orchard, as the basis for policy, generated an enabling environment for innovation, which in turn enabled greater effectiveness by reaching wider geographical areas of beneficiaries and made spatial production more accessible. While increasing distribution of productive spaces, scaling-up of innovation also meant that attention was placed on the value created through new social relationships and networks embedded in place. For urban food growing, scaling-up therefore suggested a simultaneous increase in socio-environmental consequences of a common spatial practice. Locally, this contributed to greater environmental awareness and social values, enabling the growth and replication of food growing initiatives.

Emerging citywide networks of productive spaces such as the one found in Brighton & Hove will require the capacity to support growth caused by policy. A potential barrier to social action among different groups within the neighbourhood is a lack of physical and organisational infrastructure. Scaling-up therefore needs a considered strategy of network development and combined service delivery. In this case, the coordination and linking of local organisations as delivery partners was useful in connecting to community groups to facilitate neighbourhood delivery. Through an examination of processes that permitted or restricted the scaling-up of innovation, it is shown that the policy environment and political will are important factors. The presence of a Green Party led City Council, City Sustainability Officer, and Food Partnership working in tandem were an important driving force to policy-based innovation. This coalition created space for sustainable development, stimulating policy and distribution of resources. Equally, political will was shown to be required not just by local government, but also on the part of Brighton & Hove's wider society to maintain collective pressure in formulating sustainable development policy. It was also shown however that it is necessary to query and verify local authority claims around the successes of policy-based innovations, especially how they are implemented in practice. For example, the need for greater monitoring and enforcement policy was suggested in this case by examining urban development outcomes.

On examining the reflection of socio-spatial development practice, the scaling-up of innovation helped to establish an enabling institutional context in the local authority, thereby promoting diffusion and the prompting of innovation. Diffusion was shown to expand influence beyond the urban system through the replication of planning and development policies to other cities in the UK and internationally. This caused a growth in scale and coverage of policy to further influence sustainable development. Locally, a pattern can be observed, in which one policy leads to another policy and an urban event or condition is influenced by a predecessor. For Brighton & Hove this growth is shown in the introduction of the City Food Strategy Spade to Spoon: Making the Connections (2006) and later development of the PANO6 Food Growing and Development Planning Advice Note (2011) contributing to the integration of spaces for food growing within the One Planet Framework of Sustainable Development (2013).

8.6 Interpretation of Social Innovation Pattern

This case's development is influenced by a neighbourhood initiative informing wider urban policies (Figs. 8.13 and 8.14). Identification activity begins with Brighton & Hove City Council recognising the opportunity presented by a civic development to test the integration of food growing within planning and development policy. Participatory planning across neighbourhood to urban scales contributed to the development of social relations and design of an archetype initiative. Additional identification of needs and development of relationships follows at the strategic level through collective insider lobbying, changes in local leadership and a focus towards sustainable urban development, and subsequent adoption of policy to support food growing.

To coordinate the planning and development archetype, service delivery partnerships were brokered through relationships developed with local organisations. The preceding identification and development processes led to a limited period of provision activity within the targeted neighbourhood (as well as failure to continue the initiative). Despite this setback, reflection activity leads to the scaling-up of innovation into planning policy which causes distribution of further identification opportunities for productive spaces across the city, reinforcing the cyclical nature of the proposed conceptual framework (see Chap. 4). Reflection activity also corresponds to the conceptual framework in closing the case, evident across the urban system through further adoption of planning and development policy, and nationally through shared policy learning with other local authorities to support sustainable development.

VISUAL NARRATIVE MAP LEGEND

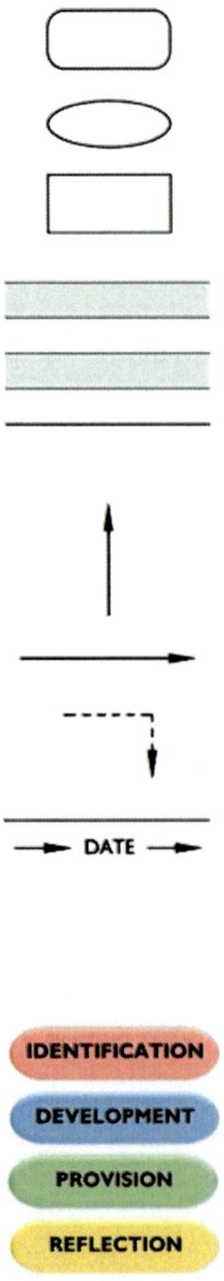

1. Polygons. The polygons on the visual narrative map describe the main elements of the case chronology relevant to social innovation.

- Rounded rectangles represent activities undertaken by the focal stakeholder
- Ovals represent events undertaken by the focal stakeholder
- Rectangles represent events and decisions outside the control of the focal stakeholder

2. Four horizontal stripes. Polygons are situated in relation to the contextual scale in which case activities and events are undertaken. Contextual scales are represented by the ascending horizontal stripes on the visual narrative map of: Neighbourhood, Urban System, and National System. The horizontal stripe to which all boxes are connected contains the base line activity of the core level of focal stakeholders that is indicated in bold.

3. Vertical arrows. Vertical arrows leading from one contextual scale to another indicate direct influences from events and activities on subsequent events and activities.

4. Horizontal arrows. Horizontal arrows within contextual scales indicate continuity of event or activity over time.

5. Dotted arrows. Indicate extended periods of time elapsing within the activity.

6. Time scales. The time scales along the bottom of each visual map vary in scale to minimise the size of maps whilst emphasising significant periods of activity and events.

7. Key social innovation processes. Key processes are represented by colour-coded polygons and situated next to the events and activities they relate to following case analysis according to the study protocols.

- Identification of human needs or societal challenges to sustainable development
- Development of social relations in systems or structures
- Provision of opportunity for social empowerment
- Reflection of socio-spatial development practice

The visual narrative maps can be read as an overview of the main events and activities within the case narrative and represent the complete case study incorporating narrative subplots. They provide a visual summary of the case study and through signifying key processes a graphical sequence from which sequence analysis can be undertaken.

(*Source* Author)

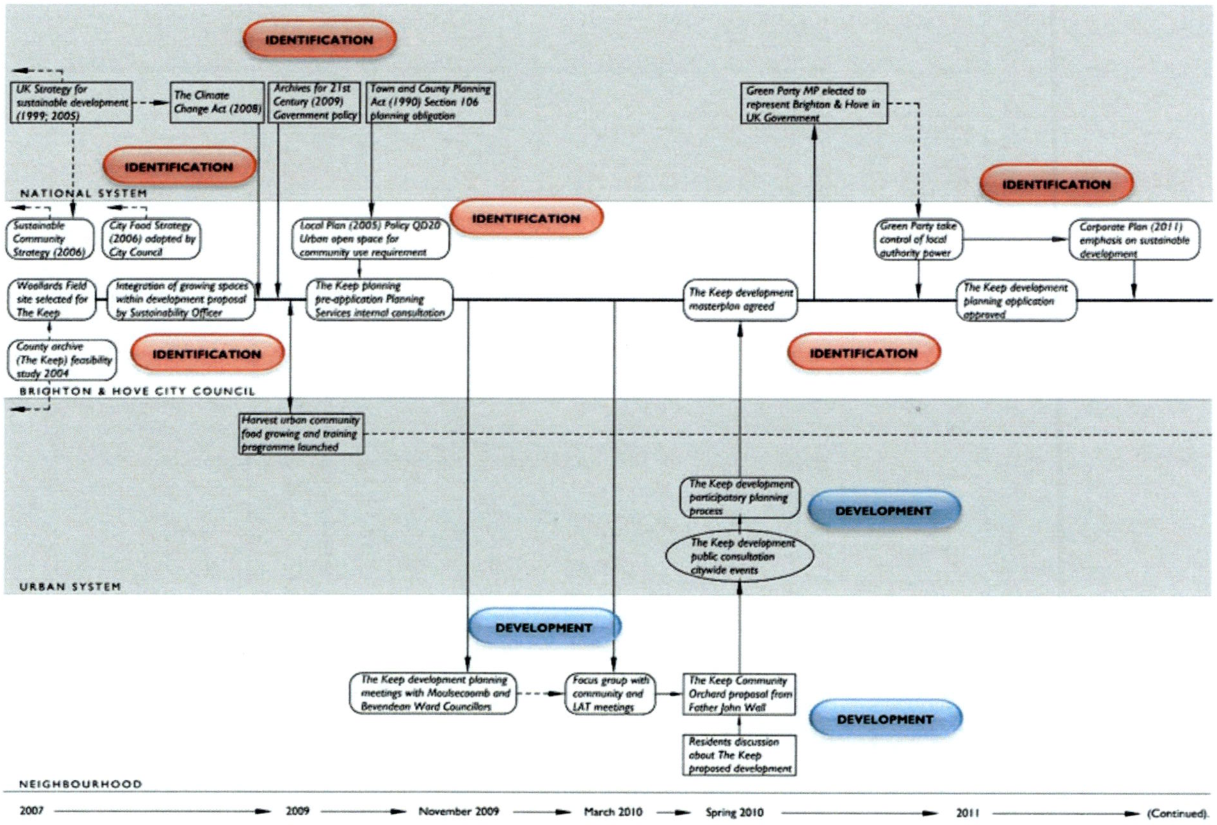

Fig. 8.13 Visual narrative map of The Keep Orchard development with key social innovation processes indicated across contextual scales. Map sheet 1 of 2 (*Source* Author)

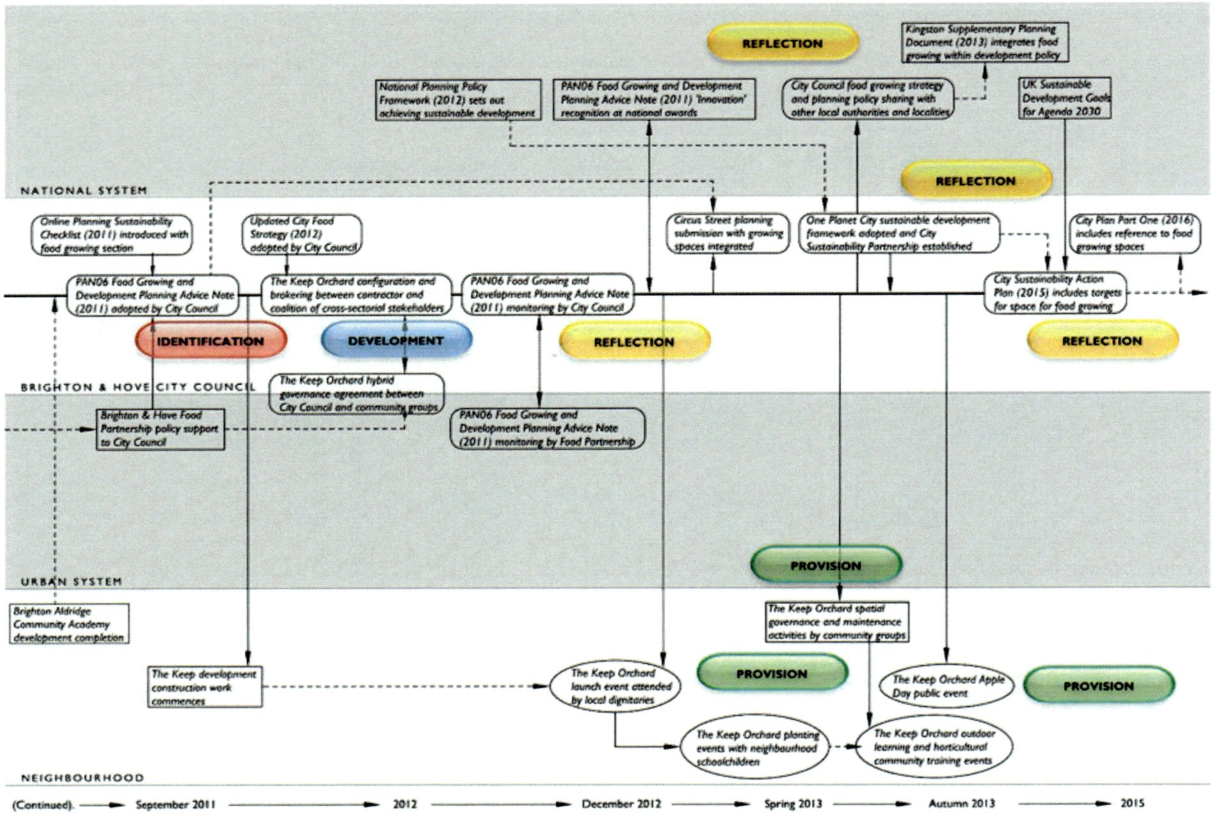

Fig. 8.14 Visual narrative map of The Keep Orchard development with key social innovation processes indicated across contextual scales. Map sheet 2 of 2 (*Source* Author)

Table 8.1 provides a summary of contextual influences across spatial scales on development.

Table 8.1 Summation of contextual influences on The Keep Orchard development

Contextual scales	Influences on project development
National system (Macro scale)	Town and Country Planning Act (1990) Sect. 106 agreement, planning obligation, set requirement for community space; UK Strategy for Sustainable Development (1999; 2005); influenced local policy; The Climate Change Act (2008), targeted the reduction of UK carbon emissions by 80% by 2050, country's food system highlighted as a significant area to be addressed; One Planet Living Framework of Sustainable Development (2013), core principles incorporated into Brighton and Hove Sustainability Action Plan (2015); Archives for 21st Century (2009) policy, influenced project brief;
Urban system (Meso scale)	Green Party elected into local authority power in 2011, advanced sustainable development policy strategies and adoption; Planning Advice Note on Food Growing and development (2011), integrated cultivation spaces within development; Brighton & Hove Food Strategy (2006; 2012), set target to make urban food growing practices more visible and to increase amount of food grown locally, informed Brighton and Hove Sustainability Action Plan (2015); Local Plan (2005) Policy QD20 for urban open space, influenced development masterplan; Public consultation process, set conditions for community stakeholder engagement; Planning masterplan negotiation, consent required;
Neighbourhood (Micro scale)	Area of multiple deprivation, social need for public facility and services to support community development; Historical land use and fruit 'scrumping' activities, informed resident brief; Neighbourhood focus group with community stakeholders, created opportunity to dictate programmatic brief; Interest in food growing practices, neighbourhood organisation with previous experience of community food growing spaces; Ceasing of initiative, affected public accessibility to green space

(*Source* Author)

References

Allard P, McSweeney J (2010) The KEEP a new historical resource centre for East Sussex and Brighton & Hove planning statement. Planning Perspectives LLP, London

Bioregional (2017) One planet goals and guidance for communities and destinations. Bioregional, London

Brighton and Hove City Council (2020) PAN 06 food growing and development. Brighton & Hove City Council, Brighton & Hove

Brighton & Hove City Council (2005) Brighton & Hove local plan 2005. Brighton & Hove City Council, Brighton & Hove

Brighton & Hove City Council (2011a) Brighton & Hove City council corporate plan 2011a–2015. Brighton & Hove City Council, Brighton & Hove

Brighton & Hove City Council (2011b) PAN 06 food growing and development. Brighton & Hove City Council, Brighton & Hove

Brighton & Hove City Council (2015) Brighton & Hove's sustainability action plan. Brighton & Hove City Council, Brighton & Hove

Brighton & Hove City Council (2016) Brighton & Hove City plan part one. Brighton & Hove City Council, Brighton & Hove

Brighton & Hove Food Partnership (2006) Spade to spoon: making the connections. Brighton & Hove Food Partnership, Brighton & Hove

Brighton & Hove Food Partnership (2014) Harvest Brighton & Hove evaluation summary 2009–2013. Brighton & Hove Food Partnership, Brighton & Hove

Brighton & Hove Food Partnership (2015) Racehill community orchard report august 2015. Brighton & Hove Food Partnership, Brighton & Hove

Brighton Permaculture Trust (2012) New orchard planned for Woollard's Field. www.brightonpermaculture.org.uk. Accessed 7 Mar 2017

Brighton Permaculture Trust (2018) Apple day brighton. https://brightonpermaculture.org.uk/apple-day-brighton/. Accessed 1 Aug 2018

Cabinet Office Strategy Unit (2008) Food matters towards a strategy for the 21st century. Cabinet Office, London

Carter W (2012) Moulsecoomb forest garden & Wildlife project. https://www.moulsecoombforestgarden.blogspot.com/2012/08/. Accessed 24 May 2019

Department for Communities and Local Government (2015) Indices of deprivation 2015: index of multiple deprivation and domain indices. Department for Communities and Local Government, London

DETR (1999) A better quality of life—a strategy for sustainable development for the United Kingdom. Department of the Environment, Transport and the Regions (DETR), London

Fainstein SS (2010) The just city. Cornell University Press, New York

HM Government (2008) Climate change act 2008. HM Govermnent, London

HM Government (2009) Archives for the 21st century. HM Government, London

Iliffe F (2014) Food growing planning advice note sustainable food cities. Brighton & Hove City Council, Brighton & Hove

Jordan H (2012) Amenity—enabling communities to grow their own. In: Horticulture-Week. https://www.hortweek.com/amenity-enabling-communities-grow-own/fresh-produce/article/1111296, Accessed 22 Feb 2019

Joss S (2015) Sustainable cities: governing for Urban innovation. Palgrave Macmillan, London

Low S, Iveson K (2016) Propositions for more just urban public spaces. City 20(1):10–31

Marcuse P (2009) From critical urban theory to the right to the city. City 13(2–3):185–197

Morriss D (2010) The keep a new historical resource centre for East Sussex and Brighton & Hove design and access statement. Atkins, Oxford

OECD (2006) Background document on public consultation. OECD, Paris

Regional K (2010) The keep a new historical resource centre for East Sussex and Brighton & Hove statement of community involvement. Kier Regional, London

Ordnance Survey (1937) Ordnance survey map of Great Britain—1:25,000. Ordnance Survey, Southampton

Sustain (2014) Planning sustainable cities for community food growing. Sustain, London

Chapter 9
Case Study of Social Innovation in Urban Food Spaces: 'Southsea Greenhouse', Southsea Greenhouse Cooperative, Portsmouth

Abstract From 2011 the Southsea Greenhouse Cooperative operated 'Southsea Greenhouse' in the city of Portsmouth, UK. This resident-led social enterprise promotes food provenance and community development, while delivering workshops to assist employability and social events to tackle social isolation. Local Southsea residents were motivated to develop this grassroots enterprise as having recognised a need for more communal facilities in the neighbourhood. Southsea Greenhouse is endorsed by multiple stakeholders in the city including Portsmouth City Council, Food Portsmouth (advocacy organisation), and several community groups. Southsea Greenhouse Cooperative's management committee comprises 12 directors who all live locally and in 2015, at its peak, the Cooperative comprised over 200 members with shared decision-making of neighbourhood space. The findings of this case indicate that an ability to change and adapt to shifting local conditions is a necessary quality to progress innovation and continue to affect urban change.

Keywords Community garden · Social enterprise · Grassroots innovation · Community development · Self-organisation · Civic participation · Social innovation · Portsmouth

Southsea Greenhouse is studied here as an experimental social enterprise in urban food growing and output of community based social innovation. The grassroots enterprise was instigated in late 2010 by Southsea resident Sue Stokes (as principal founder) with the support of Marion Dawson and Nigel Huggins and overseen by a residents' group before registering as Southsea Greenhouse Cooperative in 2012.

The following examines the Southsea Greenhouse from 2010 through to 2018. Neighbourhood development was advanced by support from Portsmouth City Council (henceforth the City Council) in providing public land for food growing in 2012.

N. Ardill, *Growing Food in Cities*, Cities and Nature,
https://doi.org/10.1007/978-3-030-98475-5_9

9.1 Identification of Human Needs or Societal Challenges to Sustainable Development

The first key social innovation process concerns being 'prompted by an experience or event or new evidence which brings to light a social need or injustice' (Murray et al. 2010, p. 14).

A principal identification activity concerned Southsea residents identifying the need to accelerate urban development as Sue Stokes, founder of Southsea Greenhouse explains: 'it was about change; the problem seemed to be nothing was happening on the seafront. If we had been visionary about investing in our city when we first had the seafront plan, a lot more would have been happening' (Int_SS). Grassroots action was prompted by the urban decline of Southsea as a tourist resort and lack of seafront development (see Fig. 9.1). In 2010, the eastern end of Southsea had few facilities, especially compared to the western end with its Historic Dockyard and Spinnaker Tower tourist areas and Gun Wharf Quays retail destination. The area suffered a decrease in footfall with several facilities closing and the main attraction of South Parade Pier fell into disrepair, not reopening until 2017. Southsea also experienced in-migration of older residents and out-migration of younger residents with the City Council recognising more services were required to meet changing local needs (Portsmouth City Council 2010).

To inform the Portsmouth Plan (2012), Portsmouth's Core Strategy, a series of planning documents were prepared. An Urban Characterisation Study (2011) identified how 'the pier and area around it have become run down and are in need of attention' (Portsmouth City Council 2011, p. 59). The Southsea Seafront Strategy 2010–2026 further highlighted:

> 'This area is dated, and its attractions are fragmented with poor facilities for visitors [...] Residents and businesses of Portsmouth wish to have better facilities at the seafront and greater use of them all year round. In order to meet their demand the City Council will need to attract more facilities and attractions to the seafront, improve what is already there, and increase the scope and number of events' (Portsmouth City Council 2010, p. 24).

As such, targeted seafront interventions were identified at bottom-up to top-down planning approaches to address lack of communal and cultural amenities in the neighbourhood. Stokes subsequently formulated a brief for a pilot enterprise and community facility that might satisfy needs, explaining:

> 'I went to the [Hampshire] School for Social Entrepreneurs because I had this burning idea that we could do something different in Southsea on the seafront. I live on the seafront. People have always complained there was not enough to do. Not enough cafes. It was about that and to test an idea. The whole idea was about doing something local in the area' (Int_SS).

Upon completion of entrepreneurial training, Stokes recruited Dawson and Huggins (via existing links) and later other participants to form a resident-led group for the proposed social enterprise.

Southsea Greenhouse's inception

Stokes's prior experience interacting with local organisations and institutions, including the City Council, influenced planning and awareness of neighbourhood

Fig. 9.1 Urban decline in east Southsea, circa 2005–2014, showing closure of South Parade Pier in state of disrepair; boarded-up facilities and community spaces around Southsea seafront and Canoe Lake Park. (Images CC BY-SA 2.0) (*Source* 'South Parade Pier', Reis 2014; 'South Parade Pier, Southsea: 2' and 'Kiosks on The Esplanade, Southsea', Jewell 2013a, b; 'Fenced off shelter on Southsea seafront', 'Zebra crossing at South Parade Pier', 'Southsea seafront', and 'Municipal Bowling Green on Southsea seafront', Eyre 2008; 'Old Portsmouth Model Yacht Club, Canoe Lake, Southsea', Baker 2005)

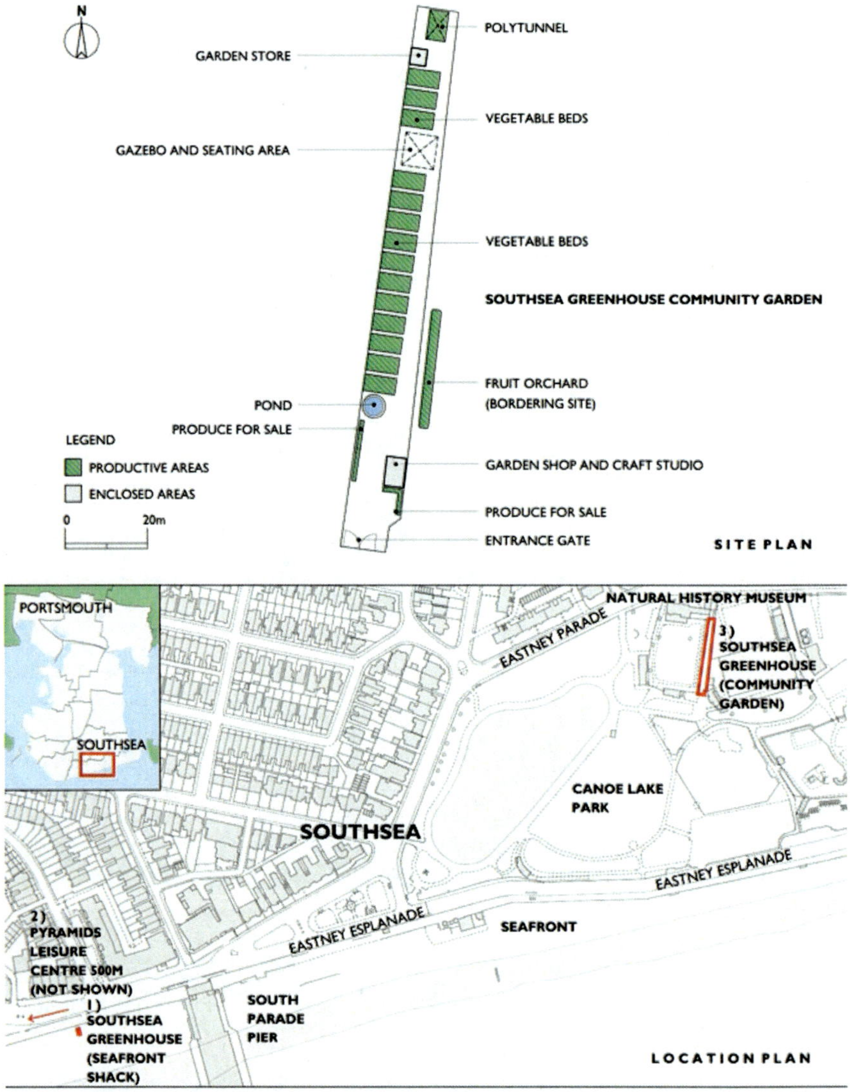

NEIGHBOURHOOD CHARACTERISATION

SOUTHSEA (EASTNEY AND CRANESWATER)

The area became urbanised as the population of the city grew and the seaside resort of Southsea expanded in the 1800-1900s. Canoe Lake is the last remnant of an area of marsh and open water drained in 1886. To the South the seafront stretches the length of the area. South Parade Pier and area around it have become run down and are in need of attention (Portsmouth City Council 2011).

Fig. 9.2 Southsea Greenhouse site and location plan (*Source* Author)
Note Three locations indicated. The enterprise's location changed over the course of urban development
In the 2015 Index of Multiple Deprivation (IMD) the Portsmouth 023D Lower Layer Super Output Area (LSOA) where the site is located was ranked 15,490 out of 32,844 in England, where 1 was the most deprived and 32,844 the least (Department for Communities and Local Government 2015).

needs. Stokes and Dawson had both worked for the Portsmouth and South East Hampshire Partnership, a non-profit urban development and regeneration organisation. The Partnership used funding from Central Government, attracting business to add value, and distributed seed funding to social enterprises to set-up or sustain. Furthermore, their 'Business Champions' programme linked business to community projects to help them be more effective:

> 'The idea was about developing capacity and skills and particularly targeting excluded and disadvantaged groups […] lots of funding to different community groups in the city, where they formed a board and got residents involved to say: we want this, we want that, and to help oversee how the money was spent. That's how I learnt how to work in the city - 10 years at working in the Partnership' (Int_SS).

Stokes also previously worked for the Royal Navy Children's Fund; creating gardens in children centres to improve spaces for Portsmouth families from high rise social housing with limited access to green space while Dawson had worked as social worker with young adults. Their combined experiences informed a social agenda to contribute to Southsea development via '3 Ps' in People, Produce, and Provenance. The strategic objectives were ambitious and wide-ranging, evolving from a premise for a neighbourhood café and florist, to later trading local produce and crafts, creating local employment, bringing together a community, reviving cultivation knowledge, personal skills development, and forming partnerships with social enterprises (Int_SS). In late 2010, aided by Stokes third sector knowhow of grant sources, seed funding of £3,000 was obtained from the national UnLtd Foundation for Social Entrepreneurs. UnLtd fund social entrepreneurship to 'transform our world for good' through innovative and sustainable business ideas that instigate social change (UnLtd 2019). This grant meant the enterprise could start looking for premises within the neighbourhood to commence operations (Fig. 9.2).

9.2 Development of Social Relations in Systems or Structures

The second key social innovation process involves the active developing of relationships and interaction between two or more individuals, groups, or organisations that create the conditions for the exchange of ideas, behaviours, and actions across multiple levels to contribute to the rearrangement and restructuring of existing social relationships (Mumford 2002; Moulaert et al. 2005).

A prominent development strategy involved residents generating cross-sector support for the initiative, initially from charities, businesses, and community groups, and then later the City Council. Stokes experience as Communications Officer for the Portsmouth Partnership was significant here to organising the development of relationships. Southsea Greenhouse Cooperative (henceforth Southsea Cooperative) utilised local media outlets, online presence, and face-to-face interactions through participating in citywide events, increasing engagement and recruitment (Fig. 9.3). This agency helped to facilitate local partnerships and collaborations to influence the case development.

Fig. 9.3 Southsea Greenhouse public engagement activities around Portsmouth (*Source* Southsea Greenhouse Cooperative 2016)

Pilot seafront 'shack': linking local producers and consumers

The residents identified vacant premises next to South Parade Pier, Southsea for the enterprise and entered negotiations with the City Council Seafront Manager, leasing from April 2011 a 3 m^2 seafront concession stand. However, the stand '[did not have] electrics or water. So, all we could really do was use it as a plant shack […] to some extent it stereotyped what we were doing. Put us in a box really, saying you are a seller of plants' (Int_SS). This restriction meant the proposed flower café was reconfigured to trading local food produce and artistic works, with the creation of a brand identity.

The enterprise launch event in April 2011 was attended by over 100 people with details of interested residents collected. Slow Food Solent (a local branch of the worldwide movement) and the University of Portsmouth provided guidance on supporting food provenance and plants were initially sourced from local charities. Trading increased through donations of produce grown by residents to the point where a waiting list was necessary and a bartering system using produce established an alternative local currency (Int_SS). Initial activity helped to form a community of food growers and meet an enterprise objective while a large chalkboard on the exterior of the shack communicated to interested passer-byers information about the enterprise and goods were displayed on the promenade (Fig. 9.4). The prominent seafront position enabled feedback and exchange of ideas with residents. Co-founder Huggins points out:

> 'We have become a real attraction on the seafront. Our aim is to buy and sell local fresh produce. We believe in real partnerships, so in addition to the Beneficial Foundation and the Shaw Trust providing plants, we have good connections with people who grow food in their gardens, small holdings and allotments […] People come along with runner and broad beans, or apples, and then swap and trade them for whatever takes their fancy' (MAST Community News 2011, para. 4).

A vegetable box delivery scheme connecting a local network of production and consumption was later established with a grant for a delivery van. However, by late 2011 the Seafront Manager sought rent increases and as the enterprise could not afford commercial business rates it became no longer viable to continue trading or pay wages. The seasonal footfall and inclement weather at the seafront was also problematic (Int_SS).

Fig. 9.4 Pilot seafront 'shack' at South Parade Pier in 2011 and Southsea Greenhouse identity (*Source* Southsea Greenhouse Cooperative 2016)

Temporary relocation and the formation of Southsea Cooperative

In early 2012 Southsea Greenhouse temporarily relocated to the Pyramids Leisure Centre, Clarence Esplanade, Southsea while exploring options for a new site. Through developing relations with the Pyramids management, an agreement was reached where space within the entrance foyer would be given over for a demonstration stall. This enabled the enterprise to continue trading and progress social innovation. Community engagement was increased through local produce and craft markets, craft activities, information about cultivation and food provenance, and operating the vegetable delivery piloted at the seafront. As such, Southsea Greenhouse continued to attempt to affect local modes of production and consumption within the urban system.

Southsea Cooperative was officially launched to the community in the spring of 2012. Cooperative membership costed £1 entitling members to contribute their opinions as to how the organisation would be co-governed and income reinvested in the social enterprise (Fig. 9.5).

During this interim period the enterprise partnered with a produce supplier to develop a distribution network and established collection hubs for vegetable boxes at the Lakeside business centre in the north of the city, the University of Portsmouth in the city centre, and the Southsea homes of its members. However, expanding the vegetable box scheme revealed challenges: 'it was just not a viable business; we found we couldn't really compete on the scale to make enough money for somebody to be employed [...] so we tried it and it has kind of failed' (Int_SS).

Fig. 9.5 Pyramids Leisure Centre, Southsea; Flyers for community events and vegetable delivery; Southsea Cooperative share selling (*Source* 'Clarence Esplanade', Sutton 2007, CC BY-SA 2.0; Additional images, Southsea Greenhouse Cooperative 2016)

Although deliveries ceased, the engagement activities at the Pyramids Centre were effective in developing community relations and expanded Southsea Cooperative's membership. Stokes says: 'what we had out of it was a group of people that were still very passionate about buying local. We were doing lots of events. We had crafters as well as gardeners coming together. So, it was about growing, buying and making local' (Int_SS). These test events later developed into 'skills workshops' for local residents when a permanent site was found. Consequently, due to lessons learned the enterprise redirected its strategic objectives around economic growth towards generating social value through delivery of community events, activities, and services.

Provision of public land to develop

Throughout 2012 Southsea Cooperative searched for a more permanent site in Southsea. At Canoe Lake (Park) they identified a bowls pavilion, closed due to public services cuts, as a potential community café and tenders were invited in early 2012 though the Cooperative could not match commercial rivals. Afterwards, through relations developed with the Friends of Cumberland House, they discovered an alternative site nearby; a 388 m^2 wasteland adjoining the Natural History Museum at Cumberland House. On this occasion the Cooperative negotiated with the Parks Services Manager (City Council) for the land, agreeing a five-year lease for a 'peppercorn rent' of 5% of any enterprise profits with business rates waived to support development in recognition of community service. The importance of this financial issue was learnt through the negative impact of business rates on the pilot seafront enterprise viability and these negotiations contrasted the Cooperative's experiences with the Sea Front Manager 'who viewed us [Southsea Greenhouse] totally differently' by not offering similar support or recognition (Int_SS). In this respect, the revitalisation and use of public land for social enterprise and communal food growing is similar to the European city case study of Bologna (see Chap. 5), although in this instance urban development was initiated at the grassroots level rather than as a local authority planning initiative.

Southsea Cooperative had to negotiate planning permission for a single storey structure to support trading and craft activities and a planning application was submitted in September 2012. Meetings were held with planning officers to negotiate the proposals prior to submission which helped to secure permission in January 2013. However, the attached conditions meant the enterprise could not meet an objective to trade as a café due to the site being restricted to selling goods associated with a community garden and for no other purposes. After securing permission, the site was cleared by volunteers ahead of a relaunch (Fig. 9.6). The site was reconfigured with raised beds, and a one-foot-square (0.1 m^2) demonstration bed to show residents how produce could be grown in a small area with limited maintenance. Southsea Cooperative also planted an orchard nearby as a legacy to the community garden

Fig. 9.6 Southsea Greenhouse development at Canoe Lake, 2012–2013 (*Source* 'The Southsea Greenhouse Community Garden project' and 'A community project', Jewell 2013b, CC BY-SA 2.0; Additional photo, Southsea Greenhouse Cooperative 2016)

if it did not continue, though due to competing land-use interests this needed to be relocated when the orchard area was later developed as commercial tennis courts. This move did however prompt the building of beneficial relations with Andy Knight, City Council Arboricultural Officer in negotiating alternative locations:

> 'I went down and looked at the trees that she [Stokes] had that she needed to move. She identified places for most of them [...] So, Wimbledon Park got a couple and we kind of horse traded over where the rest of them would go. Eventually they kind of just all got moved [...] offer them advice. It pays off in the long-term because you are being helpful, approachable and furthering their cause [helping to build up trust]' (Int_AK).

In April 2013 the community garden launch event was held and covered by local media and attended by Southsea residents, local politicians, and a representative of the Royal Horticultural Society (Fig. 9.7). Interest among the community to participate was demonstrated by 40 residents signing up as volunteers (Judd 2013).

Since opening, the community garden has connected to other local initiatives to share resources and featured in Portsmouth's first Community Gardens Open Day in July 2017 which raised awareness and encouraged food growing spaces in the city. Development of relations with organiser Food Portsmouth have been mutually beneficial as Food Portsmouth helped negotiate an extended site lease with the City Council in January 2018. Coordinator Gail Baird explains:

> 'A memorandum of understanding between us [...] support other organisations involved in promoting the growth of food in green urban spaces. It was a way we could help them if they

Fig. 9.7 Southsea Greenhouse's launch event, April 2013 (*Source* Southsea Greenhouse Cooperative 2016)

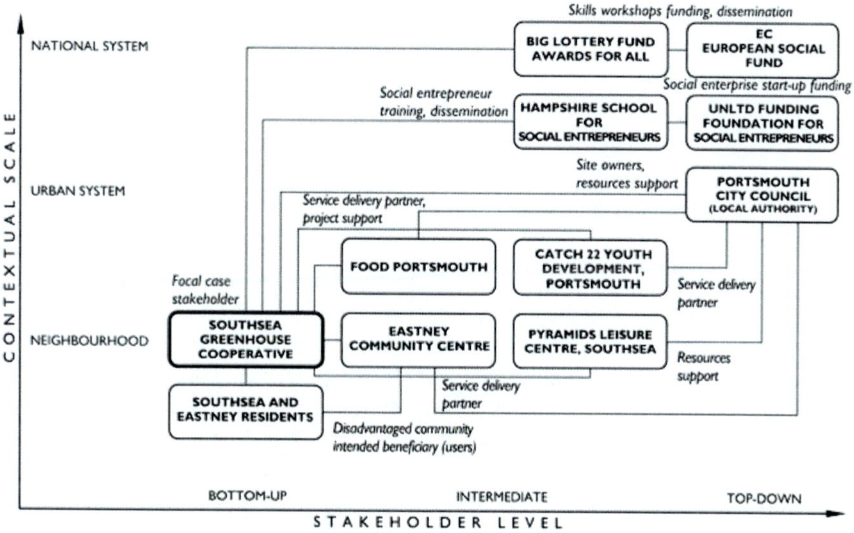

Fig. 9.8 Stakeholders and their relationships in Southsea Greenhouse development (*Source* Author) *Note* Focal case stakeholder outlined in bold

were in trouble, or if they didn't have the resources, the time or whatever [...] we'll continue to work with them. So, they'll support us in the way they can, i.e. use their network, their contacts, and volunteers' (Int_GB).

The Cooperative and Food Portsmouth collaborate on skills workshops, participate in Food Cycle Portsmouth (distributing surplus), and promote local events. Southsea Greenhouse also assist Portsmouth Friends of the Earth to highlight environmental issues and 'Green Drinks' events supporting networking, ideas and innovation. The Cooperative developed relationships with Eastney Community Centre and Catch 22 youth development charity to jointly deliver services. Additionally, they worked with the Natural History Museum to provide green activities and partnered with the Woodland Trust on their 'Plant a Million Trees' campaign, using the community garden as a base to distribute trees to community spaces around the city.

Figure 9.8 highlights key stakeholders and relationships in the Southsea Greenhouse.

9.3 Provision of Opportunity for Social Empowerment

The third key social innovation process involves the provision of opportunity for collective empowerment through increasing citizen participation and access to power and resources to meet human needs (Moulaert et al. 2005; TEPSIE 2014).

A central provision activity was the creation of participatory opportunities through community events and increased access to local food growing. The Canoe Lake site was significant in facilitating land for communal cultivation in a highly urbanised area (Food Matters 2015). In discussing this social value, Stokes points out:

> 'This land has been shared with so many more people in the city […] people come: 1) they are curious; but 2) quite often people are isolated. A lot of people who come and particularly those who want to volunteer have been lonely or older people. It has given people somewhere to meet, but also something to do. There are a lot of people who feel like they are on the rubbish heap basically. It was about having somewhere where people would come together' (Int_SS).

As the enterprise utilised resident volunteers, it also brought people together to grow a community with social opportunities for collective gardening and activities to tackle the social exclusion of segregated groups and individuals. For example, in 2015 a series of tea party events supported lonely people and those with health issues using a small grant funding of £500 from the City Council programme 'Love Your Street' that encourages residents to support projects that enhance their environment and transform their neighbourhoods. The first event had 15 attendees where gardening and craft workshops supported personal confidence building through community interaction and at the same time highlighting volunteering opportunities (Fig. 9.9). In terms of social capital, £2.16 per £1 of grant funding was generated, including nine volunteers helping, giving a total of 55 volunteer hours (if paid at basic living wage this would have cost £432). Overall, user feedback suggested a 'positive shift' in emotional state of mind (Love Your Street 2016).

Delivery of skills workshops to support community development

Southsea Greenhouse also offers training to 'local people who are unemployed or disadvantaged to improve their confidence and skills' (Social Enterprise Cluster

Fig. 9.9 Tea party community event (*Source* Southsea Greenhouse Cooperative 2016)

2015, p. 1). To build community capacities, the enterprise obtained a European Social Fund grant (a European Commission initiative to support projects with new ways of working) and delivered 'skills workshops' in food growing in small spaces, healthy cooking, crafts, mindfulness and life coaching. An inclusive programme with intergenerational events delivered 80 sessions in 2015 and 50 additional workshops were delivered to 101 participants with a £10,000 grant from the Big Lottery Fund Awards for All in 2016 (Fig. 9.10). Questionnaires monitored feedback, showing 36 people with disabilities and 39 with health issues participated from across postcode areas P01 to P16 (Southsea Greenhouse 2016). In 2018, building on previous service delivery, 30 additional workshops were funded and provided to 210 participants, showing an increased attendance. Postcodes represented included PO1 to PO10 with most from PO1 to PO4 near to the site of Southsea Greenhouse (Southsea Green 2018).

Personal and social development was supported by the enterprise working with local partners to deliver community events and social learning, especially to fulfil the objective to revive local food growing and support community development. Using £4,000 funding from Tesco supermarket they partnered with Eastney Community Centre to deliver 'Green Fingers' workshops across the two sites to bring together Eastney and Southsea communities (Fig. 9.10). The enterprise has also targeted specific demographics through delivering services. With Catch 22 they delivered vocational training to people aged 16–19, supporting personal skills and employability through alternative learning environment (see Fig. 9.11). Using £3,500 from

Fig. 9.10 Southsea Greenhouse skills workshops and Green Fingers workshops (*Source* Southsea Greenhouse Cooperative 2016)

Fig. 9.11 Southsea Greenhouse youth development workshops with Catch 22 (*Source* Southsea Greenhouse Cooperative 2016)

the City Council they delivered a 'Solent All-Stars' programme encouraging 19 participants to pitch for funding to 'bring about social change in their city'. Three of the youths subsequently gained full time employment (HIWCF 2018). Presenting an award at Portsmouth Guildhall, Penny Mordaunt, Portsmouth MP and Minister for Equalities declared the participants as being 'amazing young people doing great things for Portsmouth city' (Fatkin 2018, para. 3). Southsea Greenhouse further worked with the Portsmouth Dementia Group and local organisations supporting people with extra learning needs and people in recovery with specialised workshops and therapeutic opportunities (Int_SS).

Through local service delivery, Southsea Greenhouse has brought multiple levels of funding into the area to create employment opportunities in combination with utilising neighbourhood assets (e.g. land and labour) and stimulating active citizenship to support community development. The enterprise supported local artists, acting as a platform for them to develop their businesses while linking producers and consumers. As residents 'lease the space off the City Council, they are empowering themselves as a group in the way they want to' suggests Baird (Int_GB). Moreover, the membership structure enabled residents to participate in collective decision-making of neighbourhood space: 'There is growing membership of 200+ and it encourages a thriving network of resources and talents across the city' (Social Enterprise Cluster 2015, p. 22).

9.4 Reflection of Socio-spatial Development Practice

The fourth key social innovation process involves feedback loops for testing, learning, evaluating and refining ideas, and a range of strategies for growing and spreading an innovation (Murray et al. 2010; SIX 2010; Caulier-Grice et al. 2012; Bund et al. 2015).

An important activity involved residents undertaking reflective learning in developing their enterprise and approaches to neighbourhood development. Knowledge gained through site procurement and failure to continue the seafront enterprise influenced its reconfiguration at Canoe Lake. The secured terms of the five-year lease agreement here were more favourable in supporting development and its negotiation demonstrated the importance of communicating the utility of social enterprise with resource brokers. In 2016, the need to continuously adapt to changes in order to sustain the enterprise was emphasised by a group split attributed to intergroup differences and disagreements over increased rules. Stokes suggests: 'the people who were the managing group went overboard on a set of rules and it all became very formal and a long application form which was off putting' (Int_SS). Further conflict surrounded the group's environmental activism and campaigning and several of the management committee left. Southsea Cooperative also struggled to maintain initial volunteering generated in moving to Canoe Lake: 'the cooperative failed mostly because many of our old members signed up but then moved on and stopped sharing their time' (Star & Crescent 2016, para. 14).

Tony Keen, City Council Community Development Officer, and coordinator of the Healthy Pompey Urban Growing programme (see Chap. 11 case 'Stacey Community Orchard', Portsmouth) provides insight here into the challenges of managing community food growing initiatives:

> 'The problem with Southsea Greenhouse is they've got a lot of volunteers, but they aren't necessarily very active [...] what I've always done in community growing is that I try to raise the money to employ workers [...] it's enough to support the volunteers and keep things going. Too often this is the problem with the 'Big Society' idea is the whole reliance is on volunteers. You must have some people who are paid and designate it sometimes. [...] Yes, you've got to be democratic, yes, you've got to be consultative, but at some point, you've got to make a decision and make something happen; you know this is the problem with committee' (Int_TK).

In short, Keen indicates that successful development and continuation of community food growing spaces requires coordination within a combined governance structure of autocratic and collective-decision-making processes.

Threatened with closure, and after a period of consideration, the remaining committee members relaunched the enterprise as 'Southsea Green' in 2017. Southsea Cooperative subsequently deregistered and transitioned into the not-for-profit Southsea Green Community Garden Association in 2018 as the structure preserved membership and eligibility for grants though required less financial reporting to aid volunteer dependent enterprise.

The initiative was redesigned towards skills workshops and to overcome volunteer availability at the community garden during opening hours a different approach to custodianship was also adopted; an 'Adopt-a-Plot' initiative was launched, dissolving spatial governance and several community groups took over vegetable plots. This new governance model however meant the space offered restricted opening to the public and functioned closer to a traditional UK allotment garden association (i.e. accessible to members or invitees only) rather than offering open access to the wider public. The initiative's long-term ability to self-sustain is further challenged by Stokes, its founder and key driver, moving to Southampton, UK and being less involved in the day-to-day running since 2018.

Since launching, Southsea Greenhouse has participated in the annual Shaping Portsmouth Conference, bringing together cross-sector stakeholders to debate urban development. It featured in the Shaping Portsmouth Ambassador Programme to engage local community and business and won a Social Enterprise Award in 2013 (Shaping Portsmouth 2013). Reports made to various local-international grant bodies also distribute learning while knowledge is openly shared through online communication, interviews with media outlets, and talks given to schools and local organisations (see Fig. 9.12).

Southsea Greenhouse has contributed to socio-ecological activism in Portsmouth through spreading ideas on urban food growing, provenance, the value of public green spaces and the importance of their protection. This was shown through the group's collective social action being organised, visible and sustained over time in space. Southsea Greenhouse helped co-ordinate the #KeepPompeyGreen and #KeepMiltonGreen campaigns to tackle urbanisation, while its website instructs residents on how to object to planning applications posing a risk to the loss of green spaces. The residents also held an 'artivism' competition (i.e. art and social responsibility) on the theme of green spaces to raise awareness and the resulting artwork was displayed citywide in public exhibitions opened by Penny Mordaunt, Portsmouth MP.

In 2013, Southsea Greenhouse won a 'Gold' award in the Community Garden category of Portsmouth in Bloom horticultural competition, taking over the organisation after the City Council withdrew support in 2014 and added 'Eat my street' categories to recognise and reward local food growing. The coordination increased awareness across the city with 108 applications, a 60% rise from the previous year (Food Matters 2015). Portsmouth Friends of the Earth suggest Southsea Greenhouse 'have changed the way people think […] You have really made people think more about what is going on locally' (Int_SS). Likewise, Martin Dennison, Portsmouth and South East Hampshire Partnership suggests they represent 'a leader, a lobbyist and an activist' (Int_MD). In short, the grassroots group sought to affect the urban system and beyond through its efforts to intervene in socio-ecological change and urban development.

Overall, Southsea Greenhouse, as a demonstration space and 'social experiment' in entrepreneurship succeeded in realising its aim to test the usefulness of social enterprise (Int_SS). Experience of successes and failures were monitored, user feedback was evaluated, and learning shared at the urban to regional scale through a fellowship with the Hampshire School for Social Entrepreneurs. This activity was expanded internationally through participating in the Enspire EU (Entrepreneurial Inspiration for the European Union) programme of the European Regional Development Fund. The programme helps to transfer best practice for emulating innovation and entrepreneurship from hard to reach communities to fulfil the Europe

Fig. 9.12 Southsea Greenhouse's knowledge sharing and dissemination of learning (*Source* Southsea Greenhouse Cooperative 2016)
a Shaping Portsmouth Conference; **b** Stokes receiving an award for social enterprise; **c** Enspire EU Conference, Sweden; **d** Hampshire School for Social Entrepreneurs; **e** Talk on community business given to students at Southsea Academy; **f** Stokes giving a presentation on the Southsea Greenhouse; **g** Keep Milton Green environmental campaign in Portsmouth; **h** Southsea Greenhouse community garden during Portsmouth in Bloom horticultural competition

2020 strategy (positioning social innovation as a driver of sustainable development) (Enspire EU 2011). Southsea Greenhouse further presented at a Pan-European conference in Sweden to share development experiences (Fig. 9.12).

In terms of urban planning and development, the enterprise emerged as a Southsea facility to support the strategic aims of the Portsmouth Plan (2012) Policy PCS9 (The Seafront) to tackle urban decline and stimulate tourism by: ‘improving the quality of the open spaces […] activities that will diversify the leisure and cultural offer without detracting from the open character of the seafront’ (p. 77). Stokes points out that ‘people have come in as tourists; from France, Italians we have had in. They are surprised something happens on that kind of scale that is just open for people to come in, particularly by the sea’ (Int_SS).

An urban development ‘loop’ is also suggested by the case. This process started from an identification of needs (shared between organisational levels of city stakeholders) for neighbourhood infrastructure to support urban development, to an area

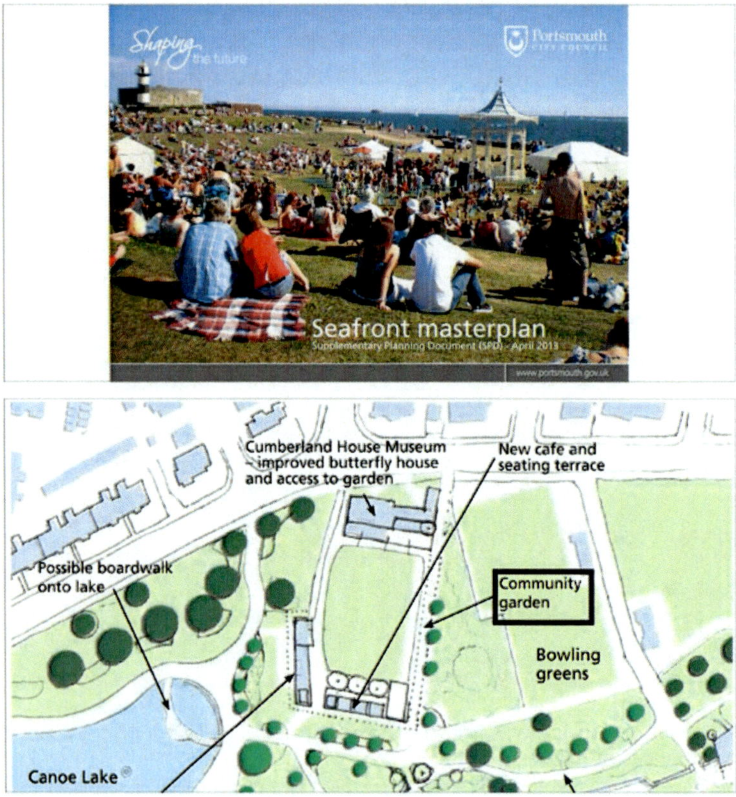

Fig. 9.13 Portsmouth Seafront Masterplan (2013) Supplementary Planning Document (SPD) urban development plan of Canoe Lake, Southsea (*Source* Portsmouth City Council 2013). *Note* Southsea Greenhouse community garden position indicated by boxed heading

of parkland being developed by local residents, and this land designated within local planning policy as a ‘community garden’. This is shown by Portsmouth’s Seafront Masterplan (2013) Supplementary Planning Document (SPD) integrating the community space within urban development plans (Fig. 9.13), which states:

> ‘To the east of the museum, on land that was under-utilised, planning permission has been granted for a community garden for the cultivation of fruit, vegetables and other plants, and for a small structure to provide storage/ a shop to sell produce grown on the site’ (Portsmouth City Council 2013, p. 69).

This masterplan builds on the Southsea Seafront Strategy 2010–2026 to provide detailed guidance about how Policy PCS9 (The Seafront) will be implemented. Additionally, Southsea Greenhouse featured as a case study and model community food garden in a draft Supplementary Planning Document (SPD) Food Growing (2015), proposing the integration of food growing spaces within urban planning (Int_TK). While this document has yet to be implemented, the adoption of similar guidance in Brighton & Hove, UK has stimulated the citywide development of productive green spaces (see Chap. 7 case ‘Racehill Community Orchard’, Brighton & Hove).

9.5 Case Discussion and Conclusions

In this section I discuss the interactions between key social innovation processes and summarise my findings. I conclude the following section with a reinterpretation of this case in key processes and spatial scales of analysis.

In summary, I analysed how a group of local residents sought to develop a neighbourhood space to support interaction and community development, and thereby address an identified need to stimulate urban development through community based social innovation. National grant funding was utilised to develop a social enterprise, piloted at the seafront, and then at a temporary location before establishing within a public park. This ‘social and physical’ neighbourhood infrastructure created social value by widening opportunities for community participation, service provision, food growing, and collective governance. Dissemination activities occurred through multiple ways in reporting and participating in events to share development knowledge.

The identification of needs for a neighbourhood social enterprise centred on prior resident knowledge of urban regeneration and community development. This suggests that bottom-up processes of identification can be configured around localised experiences and skill, especially concerning the nature of social action. Here the resident founders had a strong understanding of context and local dynamics as demonstrated in their recognition of urban decline and need for community facilities. This view is confirmed by the Southsea Seafront Strategy 2010–2026 in response to declining tourism and economic downturn (Portsmouth City Council 2010). However, these residents did not necessarily confirm this identification by undertaking any specific community consultations or participatory planning for the enterprise. This implies the enterprise’s instigators diagnosed the needs of intended social innovation beneficiaries on their behalf, rather than in conjunction with them as expected by the conceptual framework and for that reason, evidence of wider identification activity within this case is suggested as being limited. Nonetheless, through recognising the need for social entrepreneurial training and identifying the opportunity for third sector (non-profit) funding, this insight provides a key reason for activating innovation.

On examining how social relations were developed, the brokering of cross-sector support to access resources was important to progress resident-led social innovation. In particular, private sector support from a local business by the use of the of a temporary space for trading was necessary to enable the enterprise to continue when threatened with closure. It was shown that contrasting support from individuals in the public sector not only influenced the configuration of the enterprise's model, but also the approach to spatial production. The transition from a space limited to produce trading to subsequently offering community food growing can be explained through the respective failure and success of developing beneficial relationships with the local authority. Additionally, mutual aid between community groups and non-profit organisations through reciprocal exchange of resources was shown to influence the delivery of collaborative services. The contribution of civil society in supporting social innovation can therefore be energised through communication of its social value to local stakeholders and supporters. This process was shown to be augmented here through prior experience of the focal case stakeholder in working across sectors within Portsmouth and beyond to spread awareness, create engagement, and generate backing amongst various individuals, groups, and institutions for the initiative.

Regarding the provision of opportunity, my findings indicate the residents succeeded in meeting their objective to increase community participation and influence urban development in collectively developing a social enterprise. This outcome is demonstrated in membership levels, stimulation of active citizenship through volunteering, and public engagement with services. The opportunity for residents to shape urban development was enabled by the governance structure and collective decision-making processes over neighbourhood spaces. Conversely, the eventual split in their management committee is implied to have been caused by conflict over this requirement to make shared decisions, affecting subsequent delivery of services.

The establishment of a neighbourhood facility was shown to be piecemeal with opportunities for community interaction created through transformation of a seafront concessions stand, linking local food producers and consumers. However, the failure to develop this model dictated a temporary relocation and shift in focus over provision activities. This process was repeated when obtaining public land to provide access to community food growing, educational services, and social activities. Consequently, the modification to development suggests community based social innovation undergoes reconfiguration over time and that active learning plays an important role in evolution. Consequently, the ability for the enterprise to obtain grant funding to sustain and accelerate community development is explained by obtaining access to a suitable site to provide public services. The creation of employment opportunities for local service providers and the delivery of skills workshops for residents to build community capacity were significant in contributing to social empowerment. The development of collective capabilities (in terms of skills, knowledge, and decision-making powers) amongst Southsea residents here supports the idea of social innovation as contributing towards more social justice in urban spaces.

Several activities influenced the reflection of socio-spatial development practice. First, the ability of residents for undertaking social innovation is shown to be learnt on the job, through trial and error, and by adapting to local conditions to progress the enterprise. Here, adaptation implies a design sequence of 'prototyping', especially experimenting and testing in practice, to affect urban change. Related to this observation is a second assumption about bottom-up approaches to urban development. The group measured experiences and feedback of this learning in order to advance innovation with the assistance of not just existing social networks, but also with networks they themselves helped to create (e.g. links formed with other local social enterprises and community gardens). The residents understand the consequences of their actions to be prompting new behaviours within the urban system to help meet objectives for social and urban change. Third, to unlock urban resources, residents clearly understood the need for the social benefit of their enterprise to be visible, communicated, and its potential for addressing needs acknowledged by innovation brokers, including the local authority. My observation here is linked to a fourth proposition concerning socio-spatial development. It is suggested that resident-led development can be adopted within urban planning and development policy when recognised as contributing towards strategic objectives of local development frameworks such as the Portsmouth Seafront Masterplan 2013.

Lastly, how the enterprise was continuously redesigned suggests a basic relationship between key innovation processes. Through experimentation, residents learnt about developing relations and providing opportunities for empowerment. Their experiences, whether successful or unsuccessful in advancing innovation, suggested a requirement to continually redesign provision activity and enterprise objectives whilst subsequent development built upon this learning. My explanation of the case therefore draws on the changing focus of development and active learning by residents. Here, residents started with an idea and through experimentation developed a social enterprise that challenged existing community interactions with spatial production and urban development. I therefore propose that local residents are influencing the production of space and their built environment by changing existing relationships and interactions within the urban system.

9.6 Interpretation of Social Innovation Pattern

I conclude this section by reinterpreting this evolving case in relation to spatial scales of analysis and key processes to develop a pattern of social innovation (Fig. 9.14, 9.15 and 9.16). The construction of this pattern allows for a summation of contextual influences on case development (see Chap. 2). I will compare the case pattern with other cases in my cross-case analysis in Chap. 12 to identify common patterns to key processes in order to answer my research questions (see Chap. 1).

VISUAL NARRATIVE MAP LEGEND

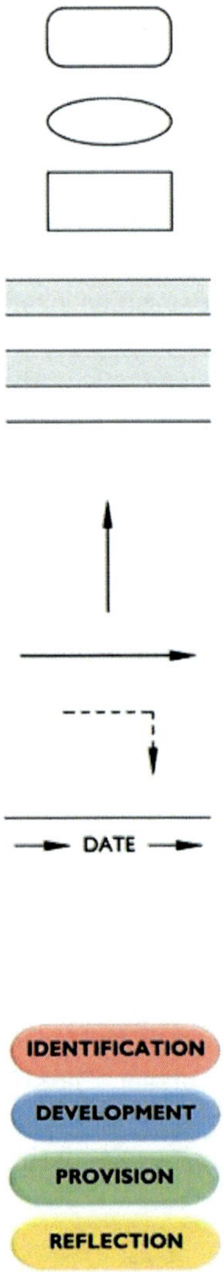

1. Polygons. The polygons on the visual narrative map describe the main elements of the case chronology relevant to social innovation.

- Rounded rectangles represent activities undertaken by the focal stakeholder
- Ovals represent events undertaken by the focal stakeholder
- Rectangles represent events and decisions outside the control of the focal stakeholder

2. Four horizontal stripes. Polygons are situated in relation to the contextual scale in which case activities and events are undertaken. Contextual scales are represented by the ascending horizontal stripes on the visual narrative map of: Neighbourhood, Urban System, and National System. The horizontal stripe to which all boxes are connected contains the base line activity of the core level of focal stakeholders that is indicated in bold.

3. Vertical arrows. Vertical arrows leading from one contextual scale to another indicate direct influences from events and activities on subsequent events and activities.

4. Horizontal arrows. Horizontal arrows within contextual scales indicate continuity of event or activity over time.

5. Dotted arrows. Indicate extended periods of time elapsing within the activity.

6. Time scales. The time scales along the bottom of each visual map vary in scale to minimise the size of maps whilst emphasising significant periods of activity and events.

7. Key social innovation processes. Key processes are represented by colour-coded polygons and situated next to the events and activities they relate to following case analysis according to the study protocols.

- Identification of human needs or societal challenges to sustainable development
- Development of social relations in systems or structures
- Provision of opportunity for social empowerment
- Reflection of socio-spatial development practice

The visual narrative maps can be read as an overview of the main events and activities within the case narrative and represent the complete case study incorporating narrative subplots. They provide a visual summary of the case study and through signifying key processes a graphical sequence from which sequence analysis can be undertaken.

(*Source* Author)

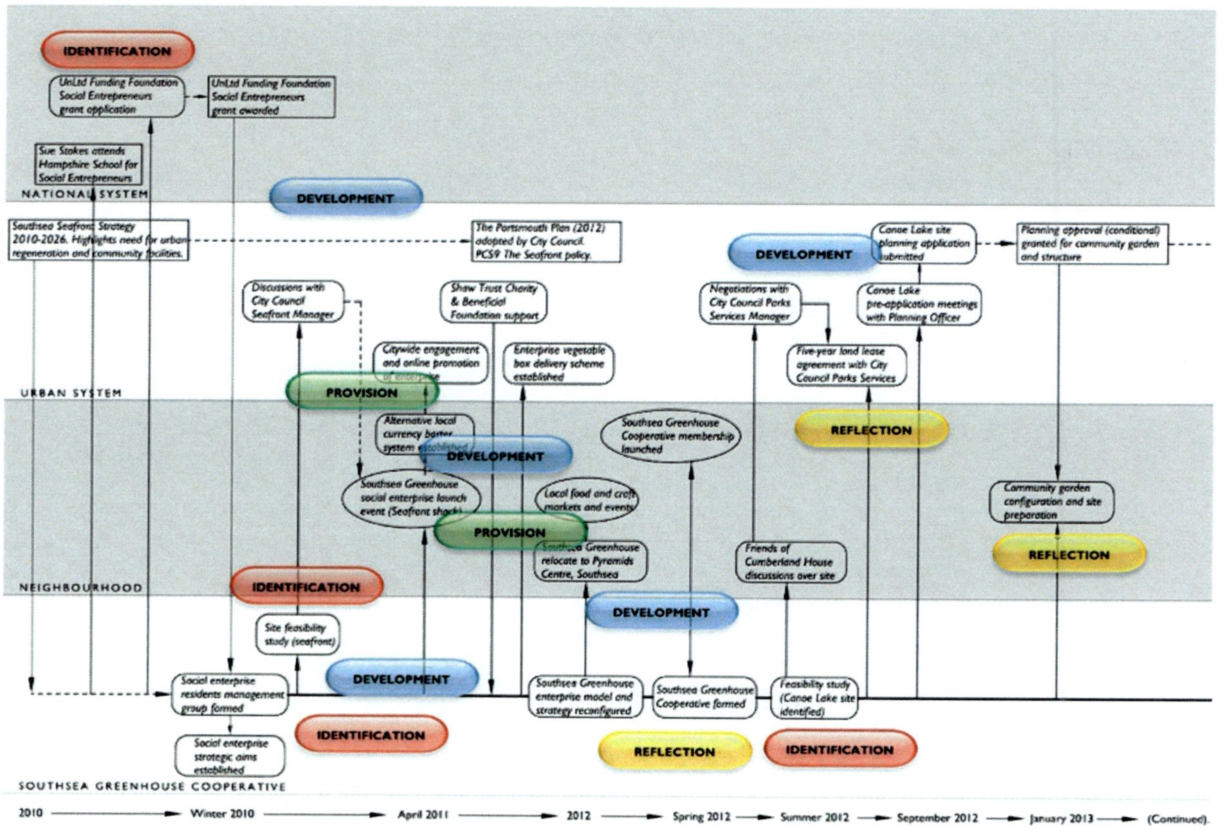

Fig. 9.14 Visual narrative map of the Southsea Greenhouse development with key social innovation processes indicated across contextual scales. Map sheet 1 of 3 (*Source* Author)

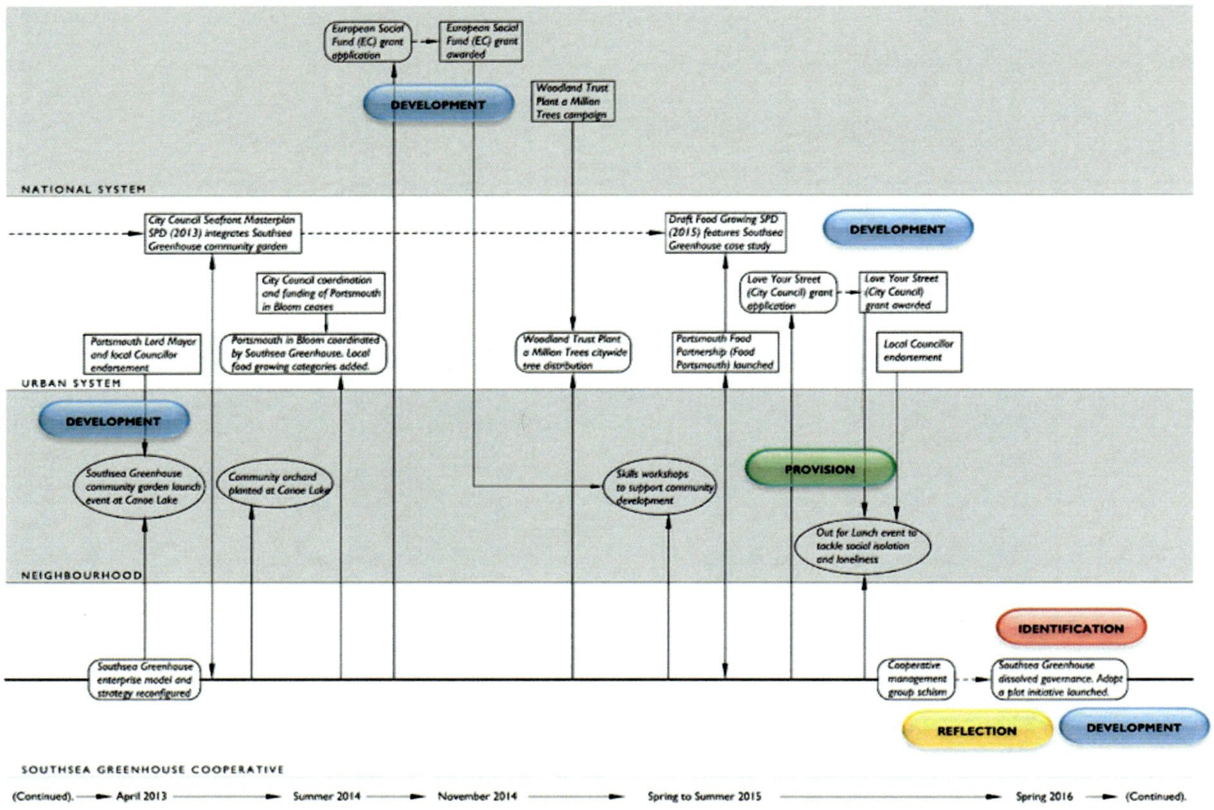

Fig. 9.15 Visual narrative map of the Southsea Greenhouse development with key social innovation processes indicated across contextual scales. Map sheet 2 of 3 (*Source* Author)

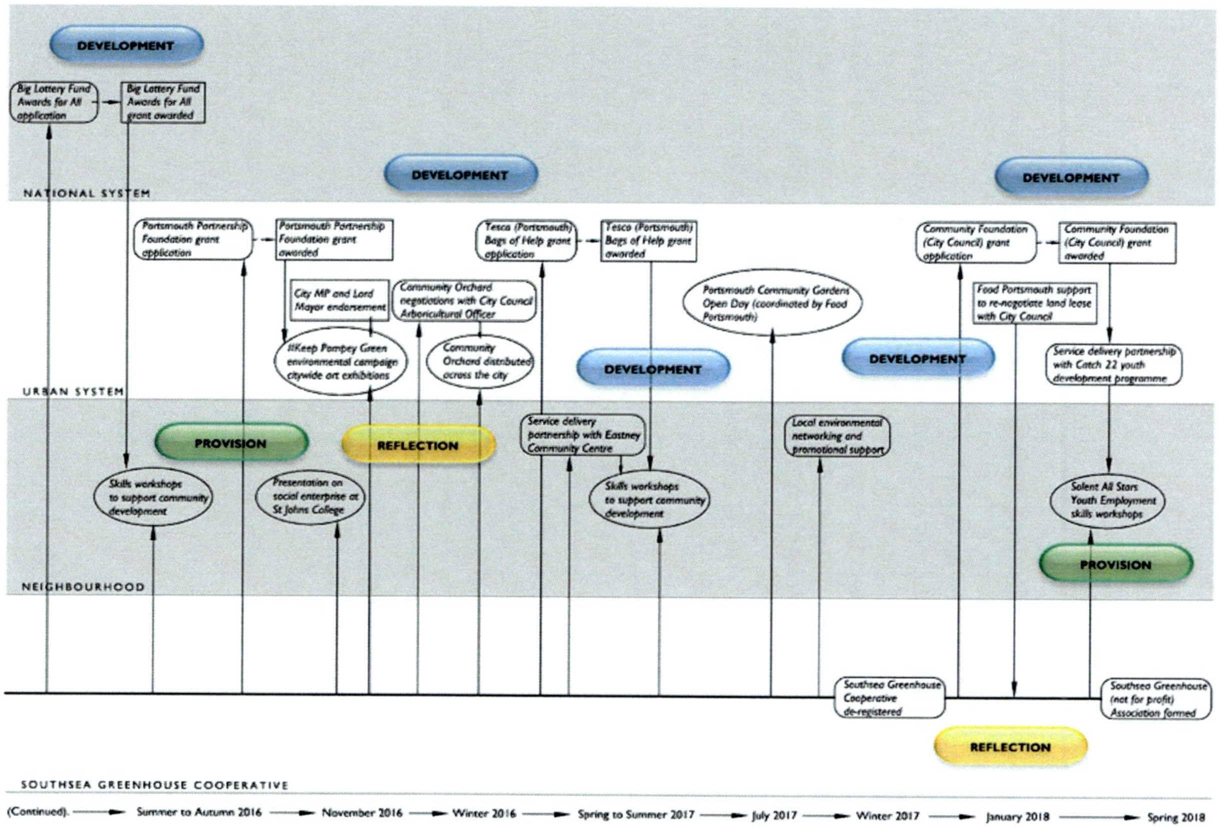

Fig. 9.16 Visual narrative map of the Southsea Greenhouse development with key social innovation processes indicated across contextual scales. Map sheet 3 of 3 (*Source* Author)

Four cycles of the proposed innovation sequence are observed across three repetitive stages of development in this case.

During the first cycle an identified need by a Southsea resident to stimulate seafront regeneration leads to the development of social relations with other neighbourhood residents to form a community cooperative, resulting in an enterprise for local produce and craft trading.

The second cycle begins with internal reflection of the group to understand the failure to sustain the enterprise model (owing to ineffective relations with the local authority as resource brokers), and to reconfigure development. The development of relations with a neighbourhood business to access material resources in space for trading enables the enterprise to relaunch at a temporary site and delivery of community events and activities begin.

The third cycle takes place over the longest period of five years, starting with the identification of opportunity for a site to enable urban food growing practices. An effective relationship with the local authority (on this occasion) creates access to material resources on favourable terms. Reflection activity is evident in learning from earlier barriers to enterprise development, and in reconfiguring enterprise aims to further opportunities. The majority of neighbourhood provision activity occurs from this time in the provision of community space for cultivation, and through the delivery of skills workshops, widening participation and supporting community development.

The fourth and final cycle is caused by disagreements over the strategic direction of the community cooperative. The resultant split necessitates further reflection of development, resulting in restricted public access to food growing spaces, focus on skills workshops, and fostering of relations with community groups to adopt cultivation plots and dissolve governance. Additional identification activity is required to source funding from multiple sources to host community activities, and relationships developed with local partners continue the service delivery. The case concluded with reflection activity in the deregistering of the community cooperative and formation of a not-for-profit association.

My analysis of Southsea Greenhouse reveals how active or ‘experiential learning’ was responsible for changes in the focus of provision over time. As such, reflecting and learning resulted in modifications to subsequent innovation agency. Learning through trial and error meant case stakeholders had to adjust and be flexible throughout their development activities as changes in the urban system occurred. The presence of experience-based learning therefore suggests a ‘feedback loop’ from development outcomes and reflection of activity to additional identification, development, and provision processes. Moreover, it demonstrates that residents are having to interact with their environment in order to solve challenges, adjust and learn. I therefore suggest that experiential learning is important to an understanding of the process of social innovation.

Table 9.1 provides a summary of contextual influences across spatial scales on development.

Table 9.1 Summation of contextual influences on Southsea Greenhouse development

Contextual scales	Influences on project development
National system (Macro scale)	• Hampshire School for Social Entrepreneurs, facilitated training; • UnLtd Funding Foundation for Social Entrepreneurs grant, enabled enterprise start-up, set guidelines to support urban development; • EC European Social Fund grant, enabled service delivery, set guidelines for community development via innovation; • Big Lottery Fund Awards for All grant, enabled service delivery, requirement for inclusive programme to support social cohesion; • Co-operative and Community Benefit Societies Act (2014), influenced governance structure;
Urban system (Meso scale)	• Portsmouth Partnership Foundation, experience in communications supported understanding of urban system dynamics and approaches to bringing development funding into the city; • City Council Parks Services, provision of public land, five-year lease for 'peppercorn rent', business rates waived supported sustainability; • Conditional planning approval granted for enterprise, pre-consultation meetings with Planning Officer, set conditions of produce trading; • Seafront Masterplan Supplementary Planning Document (SPD) 2013, recognition of Southsea Greenhouse; • City Council Arboricultural Officer, professional advice; • Food Portsmouth, supported negotiations with City Council, service delivery partnership; • Service delivery partnerships with local organisations, multiplied agency; • Love Your Street grant, set guidelines for citizen led initiatives to enhance environments and transform neighbourhoods; • Vegetable delivery scheme, failure and learning generated;
Neighbourhood (Micro scale)	• Perceived lack of opportunities (facilities) for social interaction; • Interest in community food growing and enterprise, produce donations from neighbourhood growers; • Pilot enterprise seafront shack, set guidelines for trading, failure and learning generated; • Pyramids Leisure Centre temporary relocation, supported the sustaining of enterprise; • Cooperative membership, requirement for democratic governance structure; • Group split, relaunch with 'adopt-a-plot' initiative, devolved spatial governance

(*Source* Author)

References

Baker I (2005) Geograph—photograph every grid square. https://www.geograph.org.uk/photo/71146. Accessed 13 May 2019

Bund E, Gerhard U, Hoelscher M, Mildenberger G (2015) A methodological framework for measuring social innovation. Hist Soc Res 40(3):48–78

Caulier-Grice J, Davies A, Patrick R, Norman W (2012) Defining social innovation. European Commission, DG Research, Brussels

Department for Communities and Local Government (2015) Indices of Deprivation 2015: index of multiple deprivation and domain indices. Department for Communities and Local Government, London

Enspire EU (2011) Best practices of entrepreneurial inspiration in the EU. European Regional Development Fund, Seville

Eyre B (2008) Geograph—photograph every grid square Profile for Basher Eyre. https://www.geograph.org.uk/profile/17822. Accessed 14 May 2019

Fatkin N (2018) Students rewarded for community work. In: News. www.portsmouth.co.uk/education/students-rewarded-for-community-work. Accessed 31 Jul 2019

Food Matters (2015) A review of the food system in Portsmouth. Food Matters, Brighton & Hove

HIWCF (2018) Youth social action fund. Hampshire and Isle of Wight Community Foundation, Portsmouth

Jewell K (2013a) Geograph—photograph every grid square. https://www.geograph.org.uk/photo/3321793. Accessed 14 May 2016

Jewell K (2013b) Geograph—photograph every grid square. https://www.geograph.org.uk/photo/3319374. Accessed 14 May 2016

Judd E (2013) Southsea Greenhouse launches its community garden. In: News. https://www.portsmouth.co.uk/business/southsea-greenhouse-launches-its-community-garden. Accessed 31 May 2018

Love Your Street (2016) "Love Your Street" first year review. Portsmouth Together, Portsmouth

MAST Community News (2011) Southsea Greenhouse—local food for local people. http://www.communitynewsportsmouth.co.uk/portsmouth?city?news/article/158/southsea-greenhouse-local-food-for-local-people.html. Accessed 5 Dec 2016

Moulaert F, Martinelli F, Swyngedouw E, Gonzalez S (2005) Towards alternative model(s) of local innovation. Urban Stud 42(11):1969–1990

Mumford MD (2002) Social innovation: ten cases from Benjamin Franklin. Creat Res J 14(2):253–266

Murray R, Caulier-Grice J, Mulgan G (2010) The open book of social innovation. NESTA, London

Portsmouth City Council (2010) Southsea strategy seafront 2010–2026. Portsmouth City Council, Portsmouth

Portsmouth City Council (2011) Urban characterisation study. Portsmouth City Council, Portsmouth

Portsmouth City Council (2013) Seafront masterplan Supplementary Planning Document (SPD)—April 2013. Portsmouth City Council, Portsmouth

Reis L (2014) South Parade Pier restoration petition gains support. https://www.bbc.co.uk/news/uk-england-hampshire-28960815. Accessed 14 May 2019

Shaping Portsmouth (2013) Portsmouth ambassadors. https://www.shapingportsmouth.co.uk/our-programmes/portsmouth-ambassadors. Accessed 6 Aug 2019

SIX (2010) Study on social innovation. The Young Foundation, London

Social Enterprise Cluster (2015) Social enterprise cluster prospectus 2015/16. Social Enterprise Cluster, Portsmouth

Southsea Green (2018) Awards for all: final report. Southsea Green Community Cooperative, Portsmouth

Southsea Greenhouse (2016) Awards for all skillshops—final report. Southsea Greenhouse Community Cooperative, Portsmouth

Southsea Greenhouse Cooperative (2016) Southsea Greenhouse. https://www.facebook.com/pg/SouthseaGreenhouse/photos/?tab=album&album_id=131268423599780&ref=page_internal. Accessed 8 Aug 2016

Star and Crescent (2016) Are we losing our community spirit? The struggle to keep Southsea Greenhouse alive. In: Star Crescent. https://www.starandcrescent.org.uk/2016/07/19/are-we-losing-our-community-spirit-the-struggle-to-keep-southsea-greenhouse-alive/. Accessed 1 Jul 2017

Sutton M (2007) Geograph—photograph every grid square. https://www.geograph.org.uk/photo/392420. Accessed 14 May 2019

TEPSIE (2014) Social innovation theory and research: a summary of the findings from TEPSIE. European Commission, DG Research, Brussels

UnLtd (2019) We are UnLtd. https://www.unltd.org.uk. Accessed 1 Aug 2019

Chapter 10
Case Study of Social Innovation in Urban Food Spaces: 'Charles Dickens Orchard Trail', Portsmouth and Southsea Tree Wardens, Portsmouth

Abstract Portsmouth and Southsea Tree Wardens promote the planting and conservation of trees in the city of Portsmouth, UK whilst at the same time engaging residents in community initiatives. The organisation has an outreach of 40 members and connects localised activities with broader agendas through its parent organisation the Tree Council UK who network the nationwide Tree Warden Scheme. Locally, the organisation is endorsed by multiple stakeholders and work in partnership with Portsmouth City Council, schools, and neighbourhood groups. Additionally, the Portsmouth and Southsea Tree Wardens are integrated into the city's Parks and Open Spaces Strategy 2012–2022. The Charles Dickens Orchard Trail was developed by the Tree Wardens across three stages between 2015 to 2018 with the collaboration of the Tree Council UK and Portsmouth City Council Officers being key to development. Contextually, in 2015 the housing estate surrounding Dickens Orchard included the most deprived area in the city and the 254th most deprived area in the England, placing this area inside the bottom 1% nationally. The findings of this case indicate that forming constructive state-civil society partnerships is important to progress social innovation, particularly in the design and development of public green spaces.

Keywords Community orchards · Productive urban landscape · Urban green space · Innovation intermediary · Co-production of space · Co-design · Social innovation · Portsmouth

Portsmouth and Southsea Tree Wardens were founded in 2010 by Pauline Powell, Coordinator, with funding from Southsea Town Council to enable membership to the National Tree Council. The Portsmouth and Southsea network covers the city's wards and is part of The Tree Council's UK-wide Tree Warden network, backed by the Government department, Communities and Local Government (Powell 2019).

The Charles Dickens Orchard Trail (henceforth Dickens Orchard) development was led by Dennis David, Tree Warden and is studied here as an output of practitioner based social innovation and prototype for productive green infrastructure.

The following examines Dickens Orchard from 2010 through to 2018. Urban development was advanced with support from Portsmouth City Council (henceforth

N. Ardill, *Growing Food in Cities*, Cities and Nature,
https://doi.org/10.1007/978-3-030-98475-5_10

City Council) in providing public land to cultivate and through a co-design process with a City Council Officer from 2016.

10.1 Identification of Human Needs or Societal Challenges to Sustainable Development

Identification activity here stems from a need for a national campaign to develop urban spaces to encourage people to gather and grow food. In 1999, the Tree Council UK launched ‘Hedgerow Harvest’ as part of their Hedge Tree Campaign to increase awareness of the importance of hedge trees in supporting biodiversity and to halt their decline. It also sought to reconnect residents to the heritage and seasonal practice of harvesting local produce. The initiative stimulated the development of linear fruit orchards and across the south east of England 50 were planted between 2010–2011 (Portsmouth and Southsea Tree Wardens 2011). Additionally, personal recollections of foragers were collected and compared against current experiences of European communities. An archive of conservation knowledge and skills was compiled to demonstrate the value of hedgerow food across cultures and to future generations (The Tree Council 2017).

In November 2010, during National Tree Week a demonstration fruiting hedge approximately 250 m^2 in area was planted by the newly formed Portsmouth and Southsea Tree Wardens (henceforth Tree Wardens) near to Hilsea railway station, Portsmouth (Fig. 10.1). This spatial prototype was significant in influencing the later development of similar productive spaces, especially Dickens Orchard. Prior-consultations with the City Council helped to identify the site and obtain permission to cultivate public land whilst fostering relations among urban stakeholders to increase on-going access to material resources (Int_PP). As part of the agreement, the area around the fruiting hedge was to be co-managed by the Tree Wardens on behalf of Parks Services with community groups assisting through planting and management activities while participating in training events, later becoming a local delivery model (see Chap. 5 for a European city comparison of a non-governmental organisation role

Fig. 10.1 Hilsea fruiting demonstration hedge planting, circa 2010 (*Source* Powell 2019)

in Prague and Malmö case studies). The Tree Wardens were also integrated into the city's Parks and Open Spaces Strategy 2012–2022 to fulfil 'Aim 7: To encourage and promote community involvement in the care of parks and open spaces [...] should reflect the needs of the community which uses them' (Portsmouth City Council 2012, p. 120).

The origins of Charles Dickens Orchard

Building on the Hedge Tree Campaign, the Tree Council UK launched 'Wild Hedges for Urban Edges' nationwide in 2015 to further develop productive spaces and 're-wild' public spaces. For Portsmouth, an initiative was proposed in 2016 at Cornwallis Crescent, Charles Dickens Ward where Tree Wardens aimed to transform an urban wasteland and remnant of World War II bombing into a community orchard within a deprived inner-city area. This initiative later developed into the Dickens Orchard, a linear orchard that culturally connects buildings and other landmarks associated with the writer in the city of his birth and follows the Dickens Trail developed by the Charles Dickens' Birthplace Museum, Portsmouth (see Fig. 10.2). The Tree Council UK used this cultural link to brand the initiative, attracting support from an American based donor (where Dickens visited on reading tours), and promoted it locally to increase engagement. Dickens himself was an advocate of a daily baked apple for health and historically there was an orchard between his birthplace in Landport and St. Mary's Church in Fratton where he was baptised (Dickens 2004).

The Cornwallis Crescent project was led by Tree Warden and local resident Dennis David who was concerned that recently granted permission to develop flats nearby could mean this wasteland and open space being lost to future development. Owing to the area being surrounded by social housing and highly urbanised there was an opportunity identified for this unexploited land to serve a community purpose, specifically as an orchard and vegetable garden that would increase access to fresh fruit, whilst promoting food provenance and being a legacy to the community (Int_DD). David had experience of gardening around housing blocks to enhance the urban environment and was in contact with Trish Bell, City Council Resident Development Officer (Property and Housing Services). Bell was approached with the idea to seek permission for development which involved searching records to work out who was responsible for the land and then negotiating permission with relevant City Council departments (Int_AK).

Bell and David collaborated to interpret neighbourhood needs with additional community input. Bell says: 'Consultation was more about walking around and talking to people, it was more informal than a formal one' (Int_TB). This was aided by footfall through the site, being positioned on a thoroughfare between the centre and estates. Face-to-face interactions alleviated concerns over loss of light or lack of maintenance by responding directly to the community and a barbecue attended by 60 residents was supportive:

> We had an event to talk to people, to explain to them where trees are going to be planted, the height of the trees. We did leaflets first to let residents know there was going to be an event, get responses. We got a few negative responses, though five months later we had won them over (Int_DD).

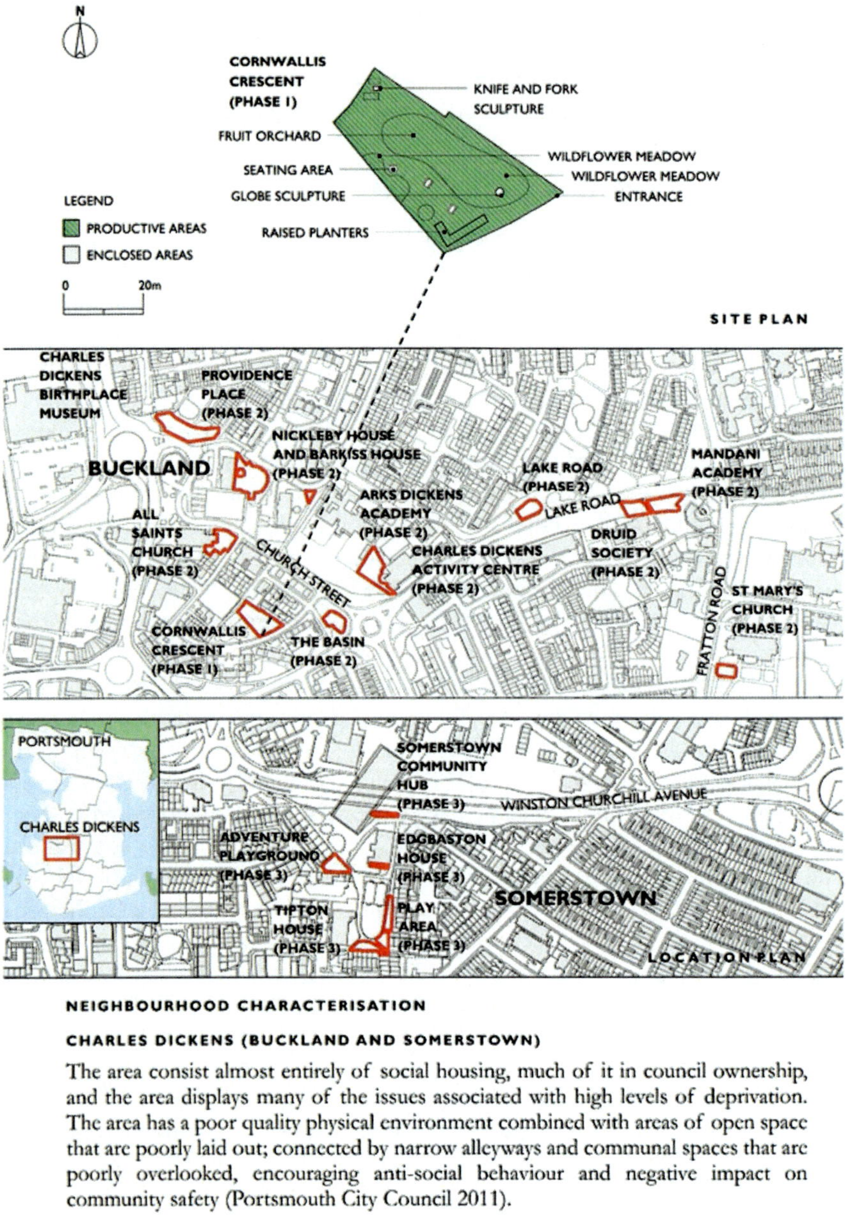

Fig. 10.2 Dickens Orchard site and location plan (*Source* Author)
Note Dickens Orchard planting phases one-three shown outlined
In the 2015 Index of Multiple Deprivation (IMD) the Portsmouth 013C and 018A Lower Layer Super Output Areas (LSOA) where the site is located was ranked 509 and 254 out of 32,844 in England, where 1 was the most deprived and 32,844 the least (Department for Communities and Local Government 2015).

Community orchards and gardens were researched in order to prepare the brief, including visiting an urban orchard in London created by the Tree Council UK, and attending an open day at the Stacey Orchard, Portsmouth to learn about local orchards (Int_TB). Stacey Orchard was evaluated by the project team as a semi-enclosed space run primarily by the City Council, whereas the proposed 'Cornwallis Community Orchard' would be unregulated and offer open access for it 'to be for everybody' (Int_DD). This 'publicness' became a key driver in development and the approach to connecting communities through collective spatial practices.

10.2 Development of Social Relations in Systems or Structures

A significant development activity concerned the use of co-production approaches in developing urban green infrastructure, particularly in the collaborative design and delivery of cultivation spaces. Collaboration between Tree Wardens, the Tree Council UK, and City Council aided project design and direction alongside implementation support from community groups and residents.

Dickens Orchard phase one: collaborative design and delivery of spaces

In progressing the Cornwallis Crescent orchard design, David met (via Bell) with Helen Bergin, City Council Landscape Architect to discuss the brief. The pair then co-designed its development phases, later becoming known as 'Dickens Orchard'. This formal translation of David's proposals into development plans by Bergin reinforced the top-down support from the City Council and also backing amongst residents, especially during the delivery phases (Fig. 10.3). Bergin explains this facilitation process as:

> I'm basically professionally advising [David], I have to try and put that [vision] into manageable language to put through to the City Council. So, there is a small tension there […] I don't want to control anything, but I have to say what's the priorities, not saying you can't do anything […] you want to get away from that telling people what they ought to do (Int_HB).

The detailed plans and perspectives helped communicate the proposal to others and meant 'we could talk to people and tell them about our plans for the area. If you wanted to apply for any grants; we had some sort of visual aid […] it is much easier to talk about things' (Int_TB). In early 2016 a small grant funding of £2,000 was obtained from local bodies to enable planting. Bell helped with funding applications in using her experience, though like Bergin, stresses project facilitation rather than steering:

> I need to have them there telling me what they want. I can help them to fill a form in but it's about what they see their vision […] its resident led, so inspired by Dennis. If they [the community] ask permission, I then must go and seek the permission, and if it's something that's within our remit we can deal with that (Int_TB).

The project team was directed by David while Bergin delivered landscape design and Bell recruited residents in forming the 'Cornwallis Community' (volunteer

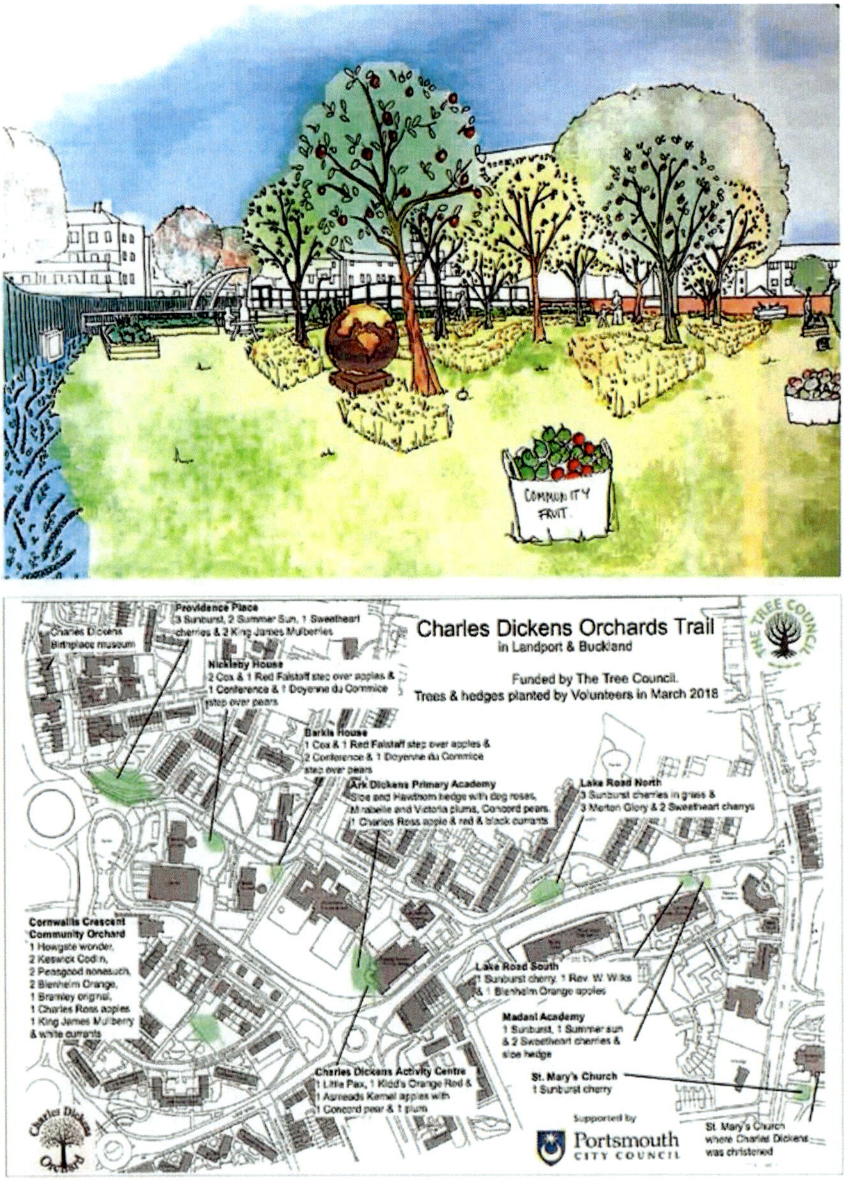

Fig. 10.3 Dickens Orchard perspective of Cornwallis Crescent and phase two planting plan (*Source* Bergin 2018a)

group) alongside other local groups to undertake tree planting. Additionally, the Tree Council UK provided planting materials and training with support provided by Andy Knight, City Council Arboricultural Officer to inform planting locations and tree management (Int_AK). Collaboration between the community and voluntary sector and public sector here influenced the case development. In particular, this support helped to activate citizens and access material resources, specifically in volunteers, project funding, and necessary consents from the local authority to develop public land.

In March 2016, the first phase of Dickens Orchard planting commenced with 31 fruit trees, numerous fruit bushes, and a large, raised vegetable bed. Volunteers from the nearby Landport Community Garden and other groups helped, forming mutual relations to enable shared resources, including use of tool stores. A launch event was held in July 2016, attended by 60 people including Councillor Stephen Morgan, Charles Dickens Ward, who was a 'keen supporter of the initiative', with the event raising awareness and promoting volunteering (Int_TB). Further planting and place-making activities were undertaken in late 2016; sculptures from the centre's shopping precinct were relocated and picnic benches installed while the sowing of a wild meadow and erection of bee houses supported biodiversity (Fig. 10.4). In terms of ongoing land management, this is jointly undertaken by Tree Wardens and residents on behalf of the City Council.

Fig. 10.4 Dickens Orchard phase one community planting and launch event in 2016 (*Source* Bell 2019)

Dickens Orchard phases two–three: growing 'an apple a day'

In early 2018, the second phase commenced, extending the Cornwallis Crescent's orchard along the Dickens Trail. Locations were proposed by the project team, comprising spaces along highways and around buildings with permission later brokered from the City Council and other local stakeholders. The city's Arboricultural Officer influenced top-down consent with David suggesting: 'you mention Andy Knight's name and it has gone a long way. Normally the City Council try to use him as a barrier, and it hasn't worked.' (Int_DD). Knight adds:

> I am prepared to talk to these people, to go out and meet people. Help them plant trees, give them advice. Previously there was a barrier […] I've built relationships with lots of people. It makes my life a little bit easier and in return their life easier to (Int_AK).

This relationship developed through Knight participating in Tree Warden activities, giving advice, and attending their meetings. The Tree Council UK, as a nationally recognised body, and the contribution of Jon Stokes, Director of Regional Programmes were also important in creating openings within the City Council to expand the initiative and thus grow social innovation: 'It's what he [Stokes] is; the key to things. The Tree Council open doors. It was a slow process until Stokes: soon as he came on board and the money to do it' (Int_DD). Meanwhile, Bell helped to coordinate community planting events, bringing together cross-sector stakeholders, and arranging publicity.

In March 2018 a public planting event was launched by the Portsmouth Lord Mayor, Councillor Kenneth Ellcome, in planting the first tree at Lake Road (near Cornwallis Crescent). Councillor Jennie Brent, Cabinet Member for Property and Housing, also present, communicated the initiative's social value:

> It's wonderful to see these hard-working volunteers giving up their time and getting involved with the orchards trail […] creating a sense of community with the added benefit of learning new skills, as well as passing on their own experience to new members. This initiative benefits so many people and will provide a long-lasting natural legacy for future generations to enjoy (Shaping Portsmouth 2018, para. 11).

The initiative was further endorsed by the Dickens Fellowship Portsmouth and planting was widespread across three days with 150 participants, encompassing residents, schoolchildren, and several community groups, including: Food Portsmouth; Abundance Portsmouth; Fratton Big Local; and Catch 22 youth development. Altogether 50 fruit trees and 400 whips (young trees) were planted (Fig. 10.5). The initiative connected additional city stakeholders in Ark Dickens Primary Academy, Charles Dickens Centre, Madani Academy (Islamic faith school), and St. Mary's Church.

Community development in the city was important to forming local relations, with Bell suggesting:

> it's a bit tribal in places Portsmouth. Having a community project brings together the community, increases the social cohesion that we really need to these days because we've got lots of cultures in Portsmouth living alongside one another (Int_TB).

David adds:

> '[Dickens Orchard] is getting bigger. It has permission to go around the city and bring other communities in, the schools, interconnecting. We share otherwise we get an isolated communities and neighbourhoods [...] I want us to be all connected and bring us together' (Int_DD).

In short, by involving others from inception, as the orchard has evolved, so have local relations between its city stakeholders and communities.

A third development phase commenced in November 2018, building on the collaborative approach of the second phase. This extended Dickens Orchard into the Somerstown neighbourhood of Charles Dickens Ward. Overall 184 fruit trees and numerous hedge plants were planted across two days around the Somerstown Community Hub and Adventure Playground sites by over 100 volunteers who came from various backgrounds (Fig. 10.6). Volunteers included Urbond, an Afro-Caribbean society who work to build community cohesion in the area, Portsmouth Friends of the Earth, Southsea Fire Brigade, and residents from local social housing blocks (Salkeld 2018).

Figure 10.7 highlights key stakeholders and relationships in the Dickens Orchard development.

Fig. 10.5 Dickens Orchard phase two community planting event, March 2018 (*Source* Author)

Fig. 10.6 Dickens Orchard news coverage and phase three planting, November 2018 (*Source* Newspapers, Melton 2018; Salkeld 2018; Photos, Bell 2019)

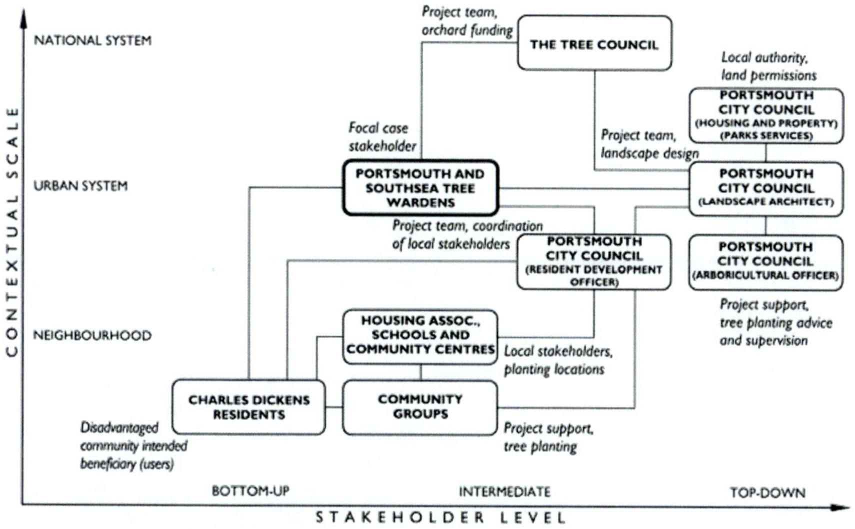

Fig. 10.7 Stakeholders and their relationships in Dickens Orchard development (*Source* Author) *Note* Focal case stakeholder outlined in bold

10.3 Provision of Opportunity for Social Empowerment

In this case, the importance of enhancing the urban environment with green infrastructure whilst bringing communities together and providing educational opportunities is evident. By developing Dickens Orchard, a direct contribution was made to transforming the urban environment and with green space limited, the initiative has widened public access. David says: 'to get half an acre in the centre of the city that is fantastic' (Int_DD). The 'urban oasis', formed between social housing blocks, is proposed to be the nearest orchard to a city centre in the UK (Int_TB). This emerging public space is certainly within 100 m from the centre's shopping precinct and provides not only environmental value, promoting quality of life in urban environment, but through transforming urban space, it offers an amenity to the community by inviting people to interact with the local area (Fig. 10.8). Moreover, it has created an identifiable place in being well-known and visited by residents. Bergin proposes: 'it has kind of created a landmark [...] When I design landscapes, if I was doing that for a design then there would definitely be focal points in there' (Int_HB). Over the last few years it has established a meeting place and a site for cultural activities, enhancing the quality of neighbourhood space (Int_TB).

The aim to unite communities through developing public green spaces for food growing was central to the provision strategy of Dickens Orchard. Stakeholder meetings enable residents to share their views and shape future development while interaction among different cultural groups is supported locally through shared activities and practices, especially cooperative tree planting, management, and community events. For example, to celebrate the anniversary of the birth of Charles Dickens a

Fig. 10.8 Dickens Orchard development of public green space at Cornwallis Crescent (*Source* Author)

planting event was held in February 2017, whilst fruit harvests as recurring events develop harvesting and knowledge among the community. By undertaking such urban encounters in a public space, residents can establish new collective identifications with one another, enhancing both the 'quantity and quality of interaction' among them (see, e.g., Low and Iveson 2016). Dickens Orchard here is the basis for the creation of a welcoming urban atmosphere and supports the development of collective culture amongst residents that is in turn influenced by the quality of urban upkeep. Moreover, Amin (2008) suggests that the building of shared aims, tolerance and care for people and places supports the production of a more 'socially just' public space.

Dickens Orchard has been accepted and cared for by residents through its appropriation by different groups at different times. David says: 'if anyone is going to mess things up, they will talk to them, not in my backyard. It's taking pride in your local area' (Int_DD). Bell adds: 'it's about having a pride in your local area; if something looks like it's well looked after, then other people are going to look after things more, but when things look run down, they are not going to look after it, they are not going to bother are they?' (Int_TB). This duty of care over the urban environment was developed through community buy-in and claiming citizens' rights over space. As such, residents:

> Begin to own it, it's our space, it's for us, it's for our future. It's them buying into it and looking after it. If they are on side with something, they want it to be there, and they want to be involved with it; they are going to look after it, especially the young people (Int_TB).

The observed 'right' to public space witnessed here connects to the arguments of Lefebvre (1996) in promoting urban civic virtue and citizenship. In this sense, the interactions of residents in an open and well-managed public space is understood to encourage tolerances, satisfaction in the urban experience, care for the shared commons, and an interest in civic life to support the development of new collective culture (Amin 2008).

A key aim of the initiative was to provide opportunities for residents within disadvantaged areas, especially among children to learn about food growing (Fig. 10.9). Local schoolchildren were considered to be increasingly disconnected from nature, the seasons and the process of food production, and the initiative would offer an opportunity to reconnect them with the natural environment through outdoor teaching (Int_TB). By involving and educating children in tree planting and ongoing care it would hopefully foster participation and continue the orchard. David suggests: 'Junior Tree Wardens. Adopt-a-tree; that's my tree, that's where I grew up. I planted that tree' (Int_DD). Bell adds: 'it's the taking part in it; that child is going to realise I planted that tree. I am going to come and keep an eye on that tree' (Int_TB). As such, the initiative may encourage children to think of future careers and spread ideas about cooking and interest in local history.

Locally, the importance of promoting the cultural link between Charles Dickens and Portsmouth to support local identity is emphasised by Councillor Donna Jones as:

> It's so important that Portsmouth is recognised as the birthplace of one of the world's most famous literary legends. This is a fantastic initiative that will see a historic trail brought to life through the hard work of dedicated volunteers. I'm sure Dickens would be proud to know that this innovative project bears his name (Shaping Portsmouth 2018, para. 6).

To create opportunities for civic participation, several local schools connect to the initiative as a result of the links Bergin had established through previous work (Int_HB). The participation of adults across neighbourhoods is also encouraged while training is given at planting events and through training sessions to support local food growing and tree care.

Fig. 10.9 Dickens Orchard planting with schoolchildren at Ark Dickens Primary Academy (*Source* Bell 2019)

10.4 Reflection of Socio-spatial Development Practice

A process of experimentation through cyclical development is evident in this case and was undertaken across three phases. The first phase at Cornwallis Crescent was significant as a design prototype to enabling the interlinked second and third phases due to embedding productive green infrastructure within the neighbourhood through demonstrating food growing to residents, whilst being accepted by the community as a neighbourhood space. Additionally, it created an argument at the strategic level to further access public resources to expand the initiative. Regarding cultural shifts in local practices, Bergin suggests:

> I think if it grows slowly, it'll probably still be here. It will have time for people to really get their heads round it rather than be a flash well here's a great idea […] if we can start looking after existing trees which is the Tree Wardens' role in the city (Int_HB).

Feedback from orchard users was obtained informally by David and Bell through face-to-face interactions with residents and through social media to share input on an ongoing basis, helping users to feel that their opinions were valued. This led to Cornwallis Crescent's space being redesigned over time to respond to resident wishes while additional feedback was gathered across planting locations for later phases, including schools and community spaces to generate new knowledge. Internal evaluation occurred after planting phases between the project team to assess successes, areas for improvements, and learning to aid future planning (Int_TB). Phase one learning was shared with local funders (via Bell) and phases two and three at the national level (via the Tree Council UK). To further share learning with environmental practitioners, the Tree Wardens attend Tree Council UK conferences and connect to the national tree warden network through visits and regional forums.

Demonstration of the initiative to support and scale-up social innovation

The assistance of City Council Officers has been central to the development and is exemplified in the coordination of local stakeholders, identification of planting locations, negotiations to obtain permissions, and orchard design. Despite this support, Bergin is eager to point out:

> It is not a council proposal; it is a resident led project. What we are trying to look at is how we continue that because it is so successful, how do we make sure it is not top level down. It is people getting interested: ownership and a feeling of contribution (Int_HB).

Significantly, Tree Warden David is positioned as the figurehead of Dickens Orchard rather than the City Council. By bringing in various neighbourhoods and communities into the initiative this supports the grassroots right of possession over public space, especially due to residents directly causing the transformation of urban space. This process is further enhanced by communicating the involvement of local people so they can identify with the initiative and therefore recording and commemorating communal planting events is important because:

> If someone takes part in something and they see a picture of them doing something, it encourages them. I took part, I did that […] to give everybody a certificate because it is the

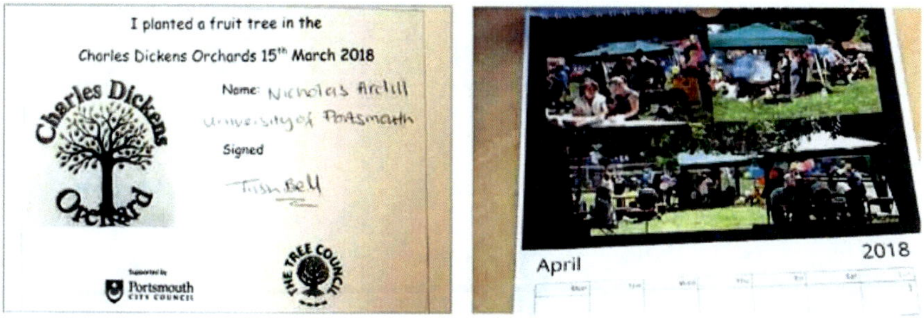

Fig. 10.10 Dickens Orchard planting certificate and calendar (*Source* Author)

> beginning of Charles Dickens Orchards. People should be led into history like that [...] I took part in planting the orchard, I planted that tree there [that connection with the landscape and the city] you are investing into the local community, your time, your energy (Int_TB).

A calendar was produced to celebrate the development (Fig. 10.10), and the City Council has recognised the work of the Tree Wardens in developing Dickens Orchard as a community resource by awarding David the 'Environment and Conservation Award' of Portsmouth Together in July 2017.

Communication to strategic level stakeholders to gain political support was important to the case trajectory and future development. Bergin explains how: 'we've got to do a three to five-year plan now to present to the councillors because it has been brought to their attention and they think it does cover a lot of [public service] directorates' (Int_HB). This masterplan identifies locations across the city wards and stakeholders who might participate, including: City Council Housing sites; parks and open spaces; schools; and private spaces such as churches (see Fig. 10.11). Presentations to the City Council and other stakeholders is led by Bergin and it is planned that David will develop these skills, 'but to start with we are giving them a professional face and confirmation that these are sensible ideas they are coming up with' (Int_HB).

Demonstration of the initiative as a low cost, low maintenance shared resource that can bring diverse communities together whilst providing a networked opportunity for healthy activities and learning is significant to accessing public resources, especially the further involvement of City Council Officers. Bell says: 'before they [City Council] invested in more support in the orchard, you need to show that you can do it, that it is working well. Because it has been working well, we've had more support to do other things' (Int_TB). There was a need to demonstrate a viable project to strategic stakeholders and involves reviewing the project plan yearly, especially land management with various departments (e.g. Education, Housing, Parks etc.), feedback from local councillors, and the Tree Wardens looking after spaces to say what has happened and where.

Development across the urban system is also reliant on the adoption and spreading of the idea among local stakeholders. To increase their involvement and generate support a consultation was undertaken, however this was acknowledged as:

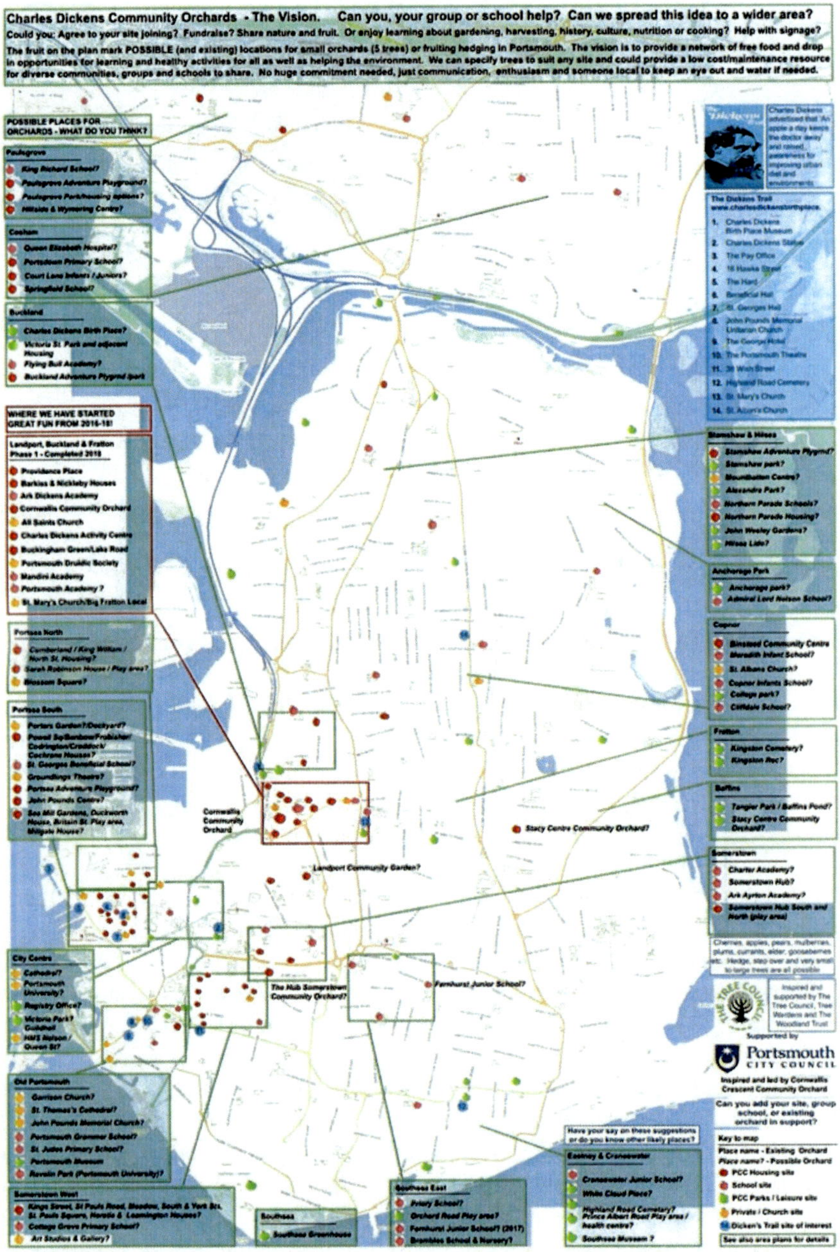

Fig. 10.11 Dickens Orchard three to five-year masterplan in Portsmouth (*Source* Bergin 2018b)

> A bit chicken and egg. We must demonstrate where we are going with it and then we double check with them, would you be happy to be [involved] I think you do come up against a bit of territorial [...] We must be very diplomatic basically. There's no reason why Stacey Centre should be affected at all by this (Int_HB).

The need for diplomacy was highlighted when the project team emailed Tony Keen, Community Development Officer and organiser of the Stacey Community Orchard, Portsmouth with advanced plans to develop the Dickens Orchard. Keen had experience of developing several community gardens and orchards for the 'Healthy Pompey' programme (see Chap. 11 case study 'Stacey Community Orchard', Portsmouth), though due to not being consulted during initial planning was:

> A bit put out by that because we are the ones with experience in developing community orchards - some of the best experience in the country and we get contacted all the time by other local authorities, but they only just contacted us now after they have pretty much developed the idea (Int_TK).

As such, the project team (and City Council) were not aware of or were unable to fully utilise local knowledge and experience in strategic planning. This raises additional questions over the potential need to retain possession or control, and even a potential 'bunker mentality' in the organisational culture of the City Council in its officers not being activity encouraged or having the opportunity to interact and communicate (Int_AK). Bell and Bergin for instance work for Housing and Property Services, whereas Keen works for another city department in Adult Social Care (Public Health). Consequently, this situation signifies a wider challenge in developing stakeholder relations, especially when undertaking organisational and citywide networking to scale-up innovation.

10.5 Case Discussion and Conclusions

In this section I discuss the interactions between key social innovation processes and summarise my findings. I conclude the following section with a reinterpretation of this case in key processes and spatial scales of analysis.

In summary, I analysed how a non-governmental environmental organisation developed a community orchard in one neighbourhood of a city followed by its enlargement across several neighbourhoods. Development was undertaken to address an identified need to enhance the urban environment whilst creating opportunities for residents to learn about orchards and local history. The Tree Wardens developed relations with individual City Council Officers to form a project team to carry the innovation forward and subsequently brought in multiple stakeholders to support its adoption. Social value was created through public planting events and the distribution of material resources across the city to affect the quality and character of urban spaces. Development knowledge was shared both internally within the project team at the city level and externally in reports given to national level funding bodies.

The identification of needs was influenced by the position of the focal Tree Warden within the community and local insight was multiplied through collaboration with the Resident Development Officer regarding her extensive place-based agency. Both of these case stakeholders resided within Charles Dickens Ward meaning that they possessed community knowledge which was gained over time and continued to grow. Being deeply ingrained in place enabled these participants to use this personal insight across subsequent development phases and innovation cycles. Local understanding was further based on their experience and the experiences of other residents and therefore embedded in community practices and relationships.

The two founding case stakeholders also considered the extent of knowledge that was already available within the local community before introducing an environmental solution. Here, gathering and integrating local knowledge was significant to using human capital as an asset in identifying needs and positioning innovation approaches to affect the built environment. Experiences and insights of residents were applied to maintain or enhance the local opportunity for wellbeing and, in order to utilise resident knowledge, appropriate participants for community consultation had to be identified. The residents with sought after knowledge were those living in proximity to the proposed spatial intervention, and consequently most likely to be affected by it as its potential users. Rather than a wider consultation, the project team conducted information gathering with certain individuals identified through informal face-to-face networks. This interpersonal agency was shown to alter during subsequent development phases across other neighbourhoods through additional City Council Officers possessing insight and local organisations representing resident opinion. As such, a process of knowledge generation and transfer was observed, through the initiative's adoption and subsequent adaption elsewhere across the urban system.

The Tree wardens, as 'change agents', had to value the insight of local residents in order to integrate community knowledge within their ongoing strategies. However, this also meant that local sources of information and knowledge needed to be critically evaluated against the nature of the project aims to determine its usefulness. Resident insight alone was not necessarily enough in developing an innovation solution and its identification is proposed as a combination of local and introduced elements. It was shown that informal insight was combined with other more formal sources, from within the urban system and elsewhere (e.g. Tree Council UK), helping to integrate insight across a variety of sources to translate needs. Also significant was how and by whom this proposed development information was communicated to both within and beyond the community. This process involved the integration of identification of urban needs at an institutional level by the City Council with national level research agendas from the Tree Council UK, to illustrate how resident insight is one component within a multi-layered innovation environment.

Key to the development of social relations was the 'gatekeeper' role of City Council Officers, especially the willingness of the Resident Development Officer and then later others to become involved. This facilitation in bridging boundaries between organisations and institutions enabled stakeholder access to public resources, both material and human with the development of relationships significant to forming a cross-sector team to undertake development. Specifically, it created the link between the Tree Wardens and residents, as users of green spaces, and the local authority as public service provider to advance the initiative. Local inhabitants as service users are implied here to be a neighbourhood resource with lived experience and therefore possess additional knowledge of community needs. From this assumption it follows that the participation of local organisations and residents has the capacity to make the delivery of urban development more effective whilst also increasing neighbourhood cohesion.

In this case, residents were no longer passive recipients of public services but active agents in changing their own environment. This dynamic signifies that a certain type of 'co-production' was occurring, one that implies a change in the traditional relationship between service user and provider. The shift in relationships over collaborative service delivery is indicated by a process in which the Tree Wardens and local residents (as end users of spaces) designed and delivered neighbourhood infrastructure in equal partnership with professionals, e.g. the City Council's Landscape Architect. By working directly together in a shared power dynamic, different kinds of knowledge and skills were combined in a reciprocal relationship. This type of cooperation is demonstrated in the mutual interactions between these two stakeholders within the built environment where each stakeholder is satisfied by the outcome. Here the Tree Wardens wanted to satisfy the need for community green spaces within the neighbourhood, whilst the Landscape Architect wanted to deliver green infrastructure that was more effective in satisfying those local needs. Relating this process to the co-production of space, it means the development of a service-resource relationship where the City Council trades a resource in the form of public land and/or professional expertise for the service of infrastructure production by a non-governmental organisation and the community who in turn get to shape their urban environment. In this sense, the building of local stakeholder networks is proposed to support more inclusive urban development processes when stakeholder relationships enable reciprocity and greater equality between participants. My observation does however question whether there is a need to reconfigure the way local government institutions design and deliver public services. In particular, to avoid isolated working and to foster robust, enduring cross-sector relationships.

In terms of the provision of opportunity, my findings indicate the Tree Wardens met an evolving objective to unite communities across different neighbourhoods in transforming the urban environment through collective action. This is demonstrated by the various groups and number of participants across multiple tree planting phases and subsequent events. These communal activities centred on the provision of public space as a site for 'collective consumption' through the collaboration of communities, and in the joint actions of orchard planting within the urban environment. This opportunity was created through the initiative being a shared and common resource among several neighbourhoods though this was shown not to be the original intent. The initiative evolved from belonging to a single location and recipients within its immediate vicinity through the opportunity created to develop and span multiple areas, and consequently be engaged by different communities. Ceremonial public events were important here in generating a collective knowledge and understanding among groups by bridging existing divides between spaces and communities. Moreover, the participants at planting events had the opportunity to interact with people they may otherwise not have met owing to different geographical and cultural backgrounds. Locally, this situation had been intensified by the city's ghettoised form (a consequence of Victorian planning, World War II bombing and post-war mass social housing), in addition to sociocultural factors, such as differences in the everyday life of neighbourhoods and communities. This collaborative process encouraged greater understanding between communities by marking the contribution they make to their shared urban environment, while also helping Portsmouth's residents to become more socially and physically connected.

The use of local media channels is implied to further communicate a message of 'urban cohesion'. Central to this process of developing resources and relationships in the built environment is a shared goal or concern among residents, such as conserving and enhancing public space, and there was an interest amongst local stakeholders in developing a community resource to support learning, including horticulture, food provenance, and local history. Consequently, this links the gathering of local information from residents, and the identification neighbourhood needs to the satisfaction of those needs. In order to continue the creation of social and spatial value through the development of community orchards, the provision of opportunities for collective action through public events such as planting, harvesting, and training was therefore necessary.

On analysing the reflection of socio-spatial development practice, the adoption and diffusion of innovation is seen to be significant to an emerging model of green space development. The novelty of introducing community orchard planting in new urban situations influenced the adoption of innovation by individuals and groups. This process was shown to involve several stages and urban participants, with the City Council's adoption of innovation central to its later diffusion among other local stakeholders and spaces. The Tree Wardens, as change agents, were responsible for introducing the innovation to individuals within the City Council who enabled it to be progressed to a decision-making level while a matching process with the City Council established knowledge of the innovation and gauged receptiveness. Here, the Tree Wardens persuaded City Council Officers as to the characteristics and suitability of the innovation to gain institutional access. Significantly, these Officers needed to possess the ability and motivation to create change, and through evaluating the advantages of the innovation, including its complexity and trialability to further confirm interest, they helped with agenda setting, consequently becoming early adopters. Therefore, the presence of 'innovation champions' within a governing institution is important to refining and supporting innovation, especially to break through potential barriers by persuading others in authority as to its social benefit.

My findings indicate that through interpersonal networks and interactions the innovation gained wider approval, meaning that decision-making responsibility shifted from the individual to a collective authority. A sequence of first nurturing the innovation in order to help it start-up, establish, and later accelerate was observed within an intentional development strategy to begin small rather than attempt to scale up too quickly and convince a wider urban population. The approach to innovation diffusion was shown to be incremental, undertaken across several implementation phases. This cyclical development involved testing Dickens Orchard to aid learning and potential adaption, whilst diffusion continued over a two-year period within the urban system and was supported by a series of communication channels. Key to innovation diffusion was the need to demonstrate that the orchard initiative was 'successful' in increasing community participation across neighbourhoods before its adoption to a city administration level to unlock further resources for growth. This success dictated a long-term masterplan for development and therefore the visible social and spatial outputs in participants and orchards developed helps top-down innovation brokers to determine its usefulness. This assessment includes an understanding of innovation application and its compatibility to satisfying local authority mandates, especially in the provision of public spaces and services to support health and education.

However, it was shown here that it is important that the consequences of innovation adoption were minimised for its adopters (e.g. City Council) and wider society through stakeholders demonstrating an ongoing commitment to management of public spaces. Furthermore, this diffusion process was shown to be reinforced by positive coverage in local media to increase public awareness. My observation here underlines the significance of internal to external channels of communication and social networks regarding both information and influence on urban change. Moreover, I suggest that the adoption of innovation by one neighbourhood helps to motivate a 'yes, in my back yard' attitude by another neighbourhood for spatial interventions.

10.6 Interpretation of Social Innovation Pattern

This case builds on that of the Southsea Greenhouse, Portsmouth (see Chap. 9) by demonstrating more than one cycle through the proposed innovation sequence (Figs. 10.12 and 10.13). However, this time it was clearly intentional due to cyclical trialling of the initiative to learn internally whilst demonstrating its social benefit to others, and thus support scaling-up of neighbourhood to urban coverage. The case closely mirrors the proposed conceptual framework (Chap. 4) in cyclical fashion from identification through to reflection, and then starting again with identification to continue a new cycle. Following preliminary activity to develop a precedent space there are two principal stages of initiative development from which two cycles of the innovation sequence can be identified. In the first cycle there is identification activity at the neighbourhood scale because of the threatened loss of open space and an opportunity identified by a local resident for this particular space to serve a greater function to the community. The relationships developed with City Council Officers brokered not only collaboration and access to public resources in land, but also funding assistance and professional services to design a community orchard initiative. Provision activity is seen in the ensuing community events and training opportunities.

VISUAL NARRATIVE MAP LEGEND

1. Polygons. The polygons on the visual narrative map describe the main elements of the case chronology relevant to social innovation.

- Rounded rectangles represent activities undertaken by the focal stakeholder
- Ovals represent events undertaken by the focal stakeholder
- Rectangles represent events and decisions outside the control of the focal stakeholder

2. Four horizontal stripes. Polygons are situated in relation to the contextual scale in which case activities and events are undertaken. Contextual scales are represented by the ascending horizontal stripes on the visual narrative map of: Neighbourhood, Urban System, and National System. The horizontal stripe to which all boxes are connected contains the base line activity of the core level of focal stakeholders that is indicated in bold.

3. Vertical arrows. Vertical arrows leading from one contextual scale to another indicate direct influences from events and activities on subsequent events and activities.

4. Horizontal arrows. Horizontal arrows within contextual scales indicate continuity of event or activity over time.

5. Dotted arrows. Indicate extended periods of time elapsing within the activity.

6. Time scales. The time scales along the bottom of each visual map vary in scale to minimise the size of maps whilst emphasising significant periods of activity and events.

7. Key social innovation processes. Key processes are represented by colour-coded polygons and situated next to the events and activities they relate to following case analysis according to the study protocols.

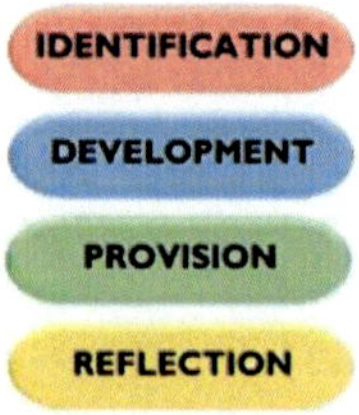

- Identification of human needs or societal challenges to sustainable development
- Development of social relations in systems or structures
- Provision of opportunity for social empowerment
- Reflection of socio-spatial development practice

The visual narrative maps can be read as an overview of the main events and activities within the case narrative and represent the complete case study incorporating narrative subplots. They provide a visual summary of the case study and through signifying key processes a graphical sequence from which sequence analysis can be undertaken.

(*Source* Author)

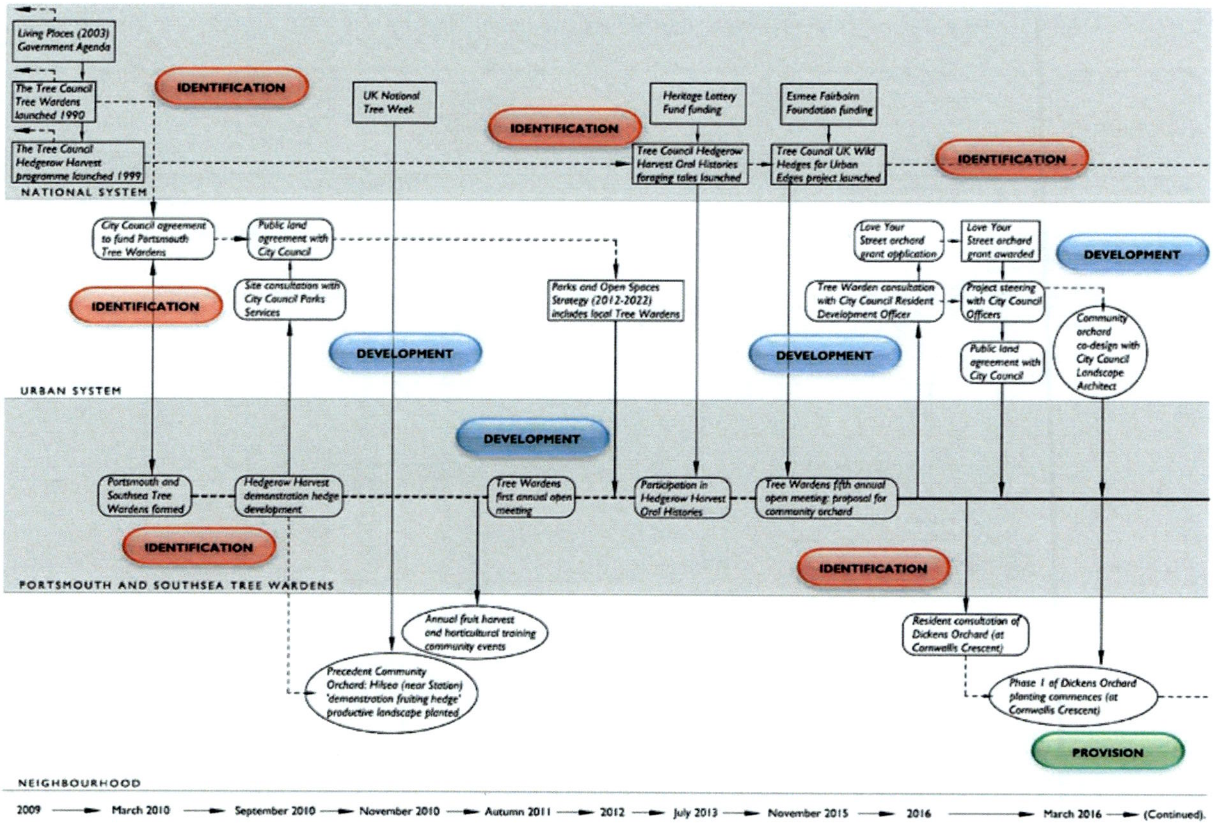

Fig. 10.12 Visual narrative map of the Dickens Orchard development with key social innovation processes indicated across scales. Map sheet 1 of 2 (*Source* Author)

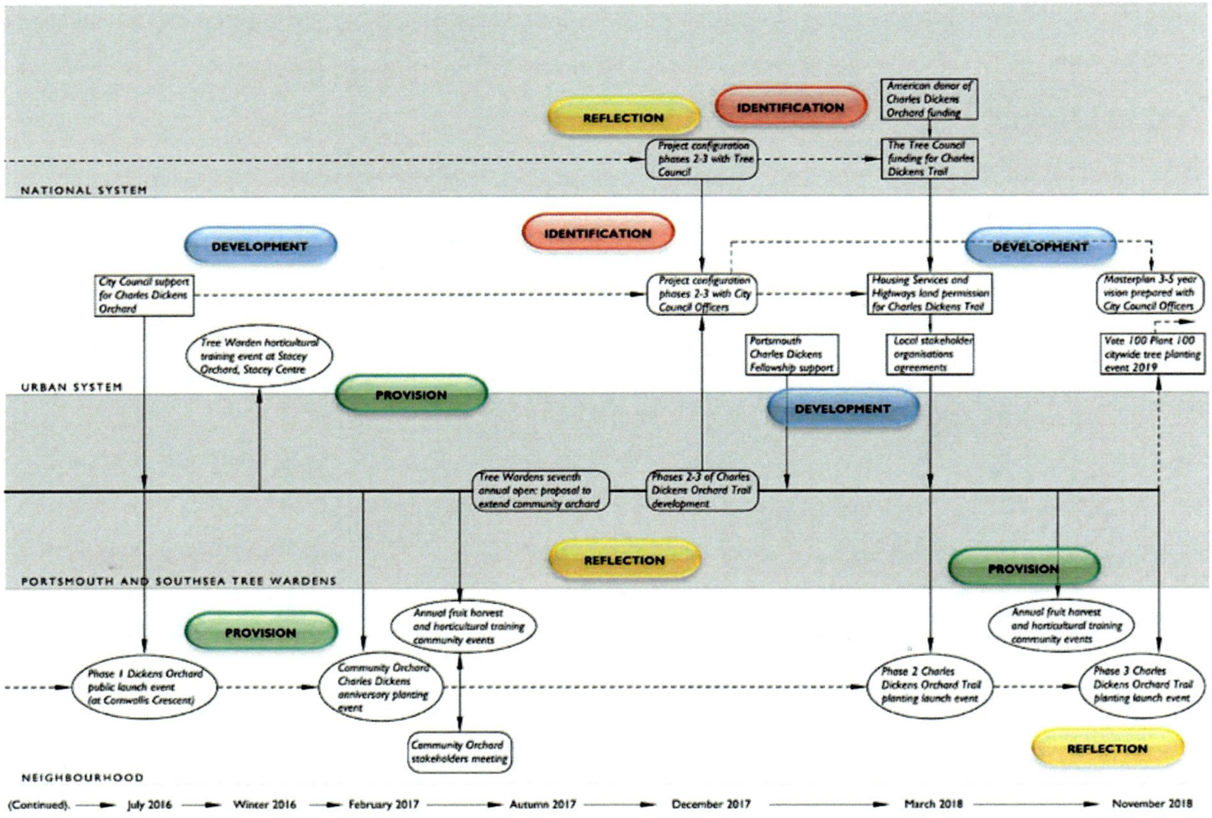

Fig. 10.13 Visual narrative map of the Dickens Orchard development with key social innovation processes indicated across scales. Map sheet 2 of 2 (*Source* Author)

The second cycle can be seen with reflection activity in evaluating the neighbourhood initiative and proposing it be enlarged and scaled-up to an urban initiative to widen social opportunities. This causes identification activity at the urban to national scales among project partners to identify resources in sites and funding, and a subsequent development of relationships with local stakeholders to broker land access for Dickens Orchard Trail. Provision activity sees multiple community planting events taking place across Portsmouth to help transform the urban environment through a collective spatial development. Consistent with the conceptual framework the case ends with reflection and learning, and then further identification activity in the preparation of an urban development masterplan setting out future growth.

Table 10.1 provides a summary of contextual influences across spatial scales on development.

Table 10.1 Summation of contextual influences on Dickens Orchard development

Contextual scales	Influences on project development
National system (Macro scale)	• Living Places: Cleaner, Safer, Greener Agenda, Government Department for Environment, Food and Rural Affairs (DEFRA) backing for Tree Warden Scheme; • Tree Council UK coordination between national and local agendas', grant funding and fruit tree provision; • Hedge Tree Campaign, Hedgerow Harvest and Wild Hedges for Urban Edges initiatives, stimulated the development of productive public spaces; • National Tree Week, nationwide event stimulates planting activities;
Urban system (Meso scale)	• Project collaboration between Tree Wardens, Tree Council UK, and City Council, cross-sector support; • Provision of public land to develop for community food growing, set conditions for spatial governance and ongoing land management; • Co-design with City Council Landscape Architect configured development phases, brokered land use permissions across City Council; • Charles Dickens' Birthplace Museum and Dickens Fellowship Portsmouth support, supported cultural and historical links and identity; • 'Love Your Street' grant, requirement for neighbourhood transformation; • Portsmouth Tree Warden network, citywide embedded member representation, awareness of local needs; • City Council Parks and Open Spaces Strategy 2012–2022 integration of Tree Wardens, community management of open spaces; • Precedent 'demonstration fruiting hedge', model for developing public spaces for community food growing, hybrid governance established;
Neighbourhood (Micro scale)	• Resident Tree Warden insight, identification of neighbourhood-based needs; • Community consultations by City Council Resident Development Officer, identification of community needs and development of programmatic brief, linked project stakeholders; • Area of multiple deprivation, high levels of urbanisation and lack of green infrastructure, unexploited public land identified for development; • Resident stakeholder meetings held to configure space and community activities, supports service delivery to users and decision-making processes

Source Author

References

Amin A (2008) Collective culture and urban public space. City 12(1):5–24

Bell P (2019) Cornwallis Crescent Community Orchard Volunteers. https://www.facebook.com/pg/Cornwallis-Crescent-Community-Orchard-Volunteers-1009344429141432/photos/?ref=page_internal. Accessed 14 Sept 2019

Bergin H (2018a) Charles Dickens Orchard Trail. Portsmouth City Council, Portsmouth

Bergin H (2018b) Charles Dickens Orchard Trail—the vision. Portsmouth City Council, Portsmouth

Department for Communities and Local Government (2015) Indices of deprivation 2015: index of multiple deprivation and domain indices. Department for Communities and Local Government, London

Dickens C (2004) The letters of Charles Dickens from 1833 To 1870. Kessinger Publishing, Whitefish

Lefebvre H (1996) Writings on the city. Blackwell Publishing, Oxford

Low S, Iveson K (2016) Propositions for more just urban public spaces. City 20(1):10–31

Melton B (2018) Legendary writer's stomping ground is set to bear fruit. In: News. https://www.portsmouth.co.uk/Legendary-writer's-stomping-ground-is-set-to-bear-fruit. Accessed 19 Mar 2018

Portsmouth and Southsea Tree Wardens (2011) Notes of the Portsmouth and Southsea Tree Wardens Training event: "How trees survive." Portsmouth and Southsea Tree Wardens, Portsmouth

Portsmouth City Council (2012) Parks and Open Spaces Strategy. Portsmouth City Council, Portsmouth

Powell P (2019) Portsmouth and Southsea Tree Wardens. https://twitter.com/portsmouthtree?lang=en. Accessed 14 Sept 2019

Salkeld M (2018) Trees planted to bear free fruit for deprived families. In: News. https://www.portsmouth.co.uk/news/trees-planted-give-free-fruit-one-portsmouths-most-deprived-areas-173954. Accessed 30 Nov 2018

Shaping Portsmouth (2018) Dickens orchard trail brings fun in spades. https://www.shapingportsmouth.co.uk/shaping-blog/1935-dickens-orchard-trail. Accessed 15 Aug 2019

The Tree Council (2017) Hedge Tree Campaign. http://www.hedgerowharvest.org.uk/About-Us/Hedge-Tree-Campaign. Accessed 22 Nov 2017

Chapter 11
Case Study of Social Innovation in Urban Food Spaces: 'Stacey Community Orchard', Portsmouth City Council, Portsmouth

Abstract Between 2008 and 2011 Portsmouth City Council participated in Change4Life, an experimental UK Government initiative to develop 'Healthy Towns'. The Portsmouth programme, 'Healthy Pompey', featured community food growing as a key component in the local authority's approach to tackle childhood obesity levels and encourage resident behavioural changes. Stacey Community Orchard was an archetype project of Healthy Pompey and demonstration garden to showcase urban food growing in small spaces. Altogether, 10 new food growing projects were developed at community spaces around the city to support health and wellbeing. Contextually, the Baffins area surrounding the Stacey Orchard had higher than average obesity levels, compared to the England average in 2008. The findings of this case study suggest that the identification and satisfaction of end-user needs through social innovation interventions can be improved if their design and delivery is more closely aligned. Future innovation strategies should ideally build in the necessary time and resources to understand local needs and environmental conditions prior to implementation.

Keywords Urban food growing · Urban development · Policy innovation · Healthy Towns · Change4Life · Behavioural changes · Social innovation · Portsmouth

The City Growing component of Healthy Pompey, including the development of Stacey Community Orchard (henceforth Stacey Orchard) was overseen by Tony Keen, City Council Community Development Officer and delivered by a team of Community Gardeners. As a local authority initiative, it is characterised as an output of policy-based social innovation (see Chap. 2).

The following examines Healthy Pompey and the development of Stacey Orchard from 2008 through to 2018. Urban development was advanced by Healthy Towns funding in 2008.

N. Ardill, *Growing Food in Cities*, Cities and Nature,
https://doi.org/10.1007/978-3-030-98475-5_11

11.1 Identification of Human Needs or Societal Challenges to Sustainable Development

Portsmouth City Council (henceforth City Council) recognised a strategic need to reduce urban obesity levels and health inequalities in relation to national concerns, as highlighted in Government reports: Tackling Obesities: Future Choices (2007) and Health Survey for England 2008 that exposed a growing problem of 'lifestyle diseases', of which obesity is the foremost. For England, 30% of children and 61% of adults were identified as overweight or obese. This 2008 figure was predicted to rise to 90% of adults by 2050 (Government Office for Science 2007), with resultant increases in medical conditions such as heart disease and diabetes estimated to pose a financial burden of £50 billion to the National Health Service (Department of Health 2008a).

Change4Life campaign and 'healthy towns' in England

In response, the Department of Health launched the Healthy Community Challenge Fund: a government-funded programme of interventions to modify the 'obesogenic environment', i.e. environmental factors affecting residents' nutrition and physical activity, in nine demonstration 'Healthy Towns', allocating £30 million between 2008 and 2011 (Department of Health 2008b). Nine urban areas experiencing high levels of obesity were selected to 'trial innovative ways of changing residents' behaviour to live healthier lives. At the launch, Alan Johnson, Secretary of State for Health (2007–09) stressed the nationwide challenge:

> Obesity is the biggest health challenge we face. For the first time we've given nine areas healthy town status. This means they must promote healthy living. But healthy towns is just the start. Our aim is to create a healthy England (Carvel 2008, para. 15).

The health programme was based upon a European model, *Ensemble Prévenons l'Obésité Des Enfants (EPODE)* which demonstrated a reduction in the socio-economic gap in obesity prevalence in France (Borys et al. 2010). The aim was to prevent obesity through a whole-town, capacity-building approach by bringing together multiple stakeholders, including education and health professionals, retailers, and the media, in a coordinated local campaign to promote active lifestyles and healthy eating initiatives. Community level interventions and local insight would support the implementation of effective and sustainable strategies (Department of Health 2008b).

The aspiring healthy towns took part in a coalition, Change4Life, that was developed as part of the Government Strategy: Healthy Weight, Healthy Lives (2008) which aimed to reduce the proportion of overweight and obese children to year 2000 levels by 2020 (HM Government 2008). Central to Change4Life's strategy in stimulating social and urban change was health-related 'social marketing' in 'the systematic application of marketing concepts and techniques, to achieve specific behavioural goals, to improve health and reduce health inequalities' (French and Blair-Stevens 2006, p. i). The campaign involved advertising, marketing, public relations, and communications aimed at local inhabitants with the advertising agency

Fig. 11.1 Change4Life brand and Healthy Pompey identity (*Source* Department of Health 2010)

M&C Saatchi creating the Change4Life brand, several sub-brands and toolkits for local level partners to use (Fig. 11.1).

Government labelling of Change4Life was avoided on the basis that research indicated that people preferred being part of a 'social movement' perceived to be owned by all, rather than prescribed by the top-down (Department of Health 2010). Change4Life was positioned by Andy Burnham, Secretary of State for Health (2009–10) as being:

> Not just a campaign but a societal movement, Change4Life, with a mission to encourage people to eat well, move more and live longer. Change4Life set out to move beyond traditional communication, with an ambitious social marketing strategy [...] achieving our shared aim – to create a lifestyle revolution in which we all play a part in changing the behaviours that can lead to people becoming overweight and obese (Department of Health 2010, p. 4).

Healthy Town submissions required local authorities to match Government funding and set out their urban strategies for encouraging healthier lifestyles. A holistic, systems-based thinking approach to innovation was encouraged in the bids to promote physical activity both in the community and in the development of urban infrastructure in densely populated built environments (see Chap. 5 for a European city comparison of strategic approaches to the development of public green spaces in the Tirana case study). Change4Life sought to especially target families with children aged 5–11 living in socially deprived areas, a demographic recognised as being 'most at risk' to becoming overweight or obese (Department of Health 2010).

Translating a national agenda to a local health programme

The City Council evidenced health needs to support Portsmouth's bid. In 2008, the city had higher childhood obesity rates than average for England: 15% of children aged 5 and 6 were overweight with 13% obese (national average 13 and 10%); and 15% of children aged 10 and 11 were overweight with 21% obese (national average 14 and 18%) (Portsmouth Primary Care Trust 2008). Reducing childhood obesity was a key health target of Portsmouth's Local Area Agreement (2008–11) between the local authority and Central Government for developing sustainable community strategies as part of The UK Sustainable Communities Act 2007 (Portsmouth City Council 2008a).

In December 2008, Portsmouth was awarded £3.1 million from the Government in combination with £2.9 million in matched funding from the local authority, totalling £6 million for Healthy Pompey of which £840,000 was allocated to community food growing projects (Portsmouth City Council 2009a). Healthy Pompey was delivered by the City Council and Portsmouth Primary Care Trust while Rachel Hatcher, Healthy Pompey Programme Manager oversaw the coordination of programmatic elements. The overarching aim was to 'make activity and healthy food choices easier for people through tackling social and environmental barriers while developing solutions to high population density and the existing transport and infrastructure network' (Brilliant Futures 2010, p. 5).

To target spatial interventions, a Health & Lifestyles Survey (2006) identified local wards with the greatest proportion of overweight children and children most 'at risk' in order to establish where the need was greatest and health gains could be made. Several deprived areas and disadvantaged communities were identified in Portsmouth, specifically families with young children within Paulsgrove, Fratton, and Baffins Wards (where Stacey Orchard is located). A social marketing survey gained further insight into the health behaviours of socio-economic groups with Hannah Byrne, Social Marketing Lead for Healthy Pompey, proposing: 'local in-depth research in three target wards increased our understanding of local families' eating habits and levels of physical activity. That insight enables us to develop specific activities and campaigns tailored to the people whose current behaviour places their children most at risk of excess weight gain' (Portsmouth City Council 2011, p. 6). The survey revealed these areas featured high instances of 'Post-Industrial Workers' and 'Blue Collar Roots' communities, indicating people living on urban estates with inactive lifestyles and low fruit and vegetable consumption (Brilliant Futures 2010).

The City Council's overall 'success criteria' for Healthy Pompey and Portsmouth stated:

- The health of Portsmouth residents is sustainably improved through cultural changes in attitude and approach. Residents feel well and are empowered to take responsibility for their health. No-one in Portsmouth will be left unaware of the Change4life and Healthy Towns messages.
- Health inequalities between areas of the city are reduced.
- Improvements to the city infrastructure in terms of growing, play, leisure, cycling, active travel and designing out obesogenic environment.
- Whole systems-change around the co-operative design and delivery of services that promote healthy lifestyles and active travel (Portsmouth City Council 2009b).

The Healthy Pompey strategy of urban change centred on key themes: Food and Health; Active Travel; City Growing; and Social Enterprise. Of interest, City Growing proposed three interlinked elements: (1) Develop a minimum of eight new community food growing spaces across the city while developing participants growing skills through training sessions and information (see Fig. 11.2 for sites); (2) Home food growing project, including distributing 5,000 growing packs to residents; (3) Identifying unused land that has potential use for community food growing (Portsmouth City Council 2010).

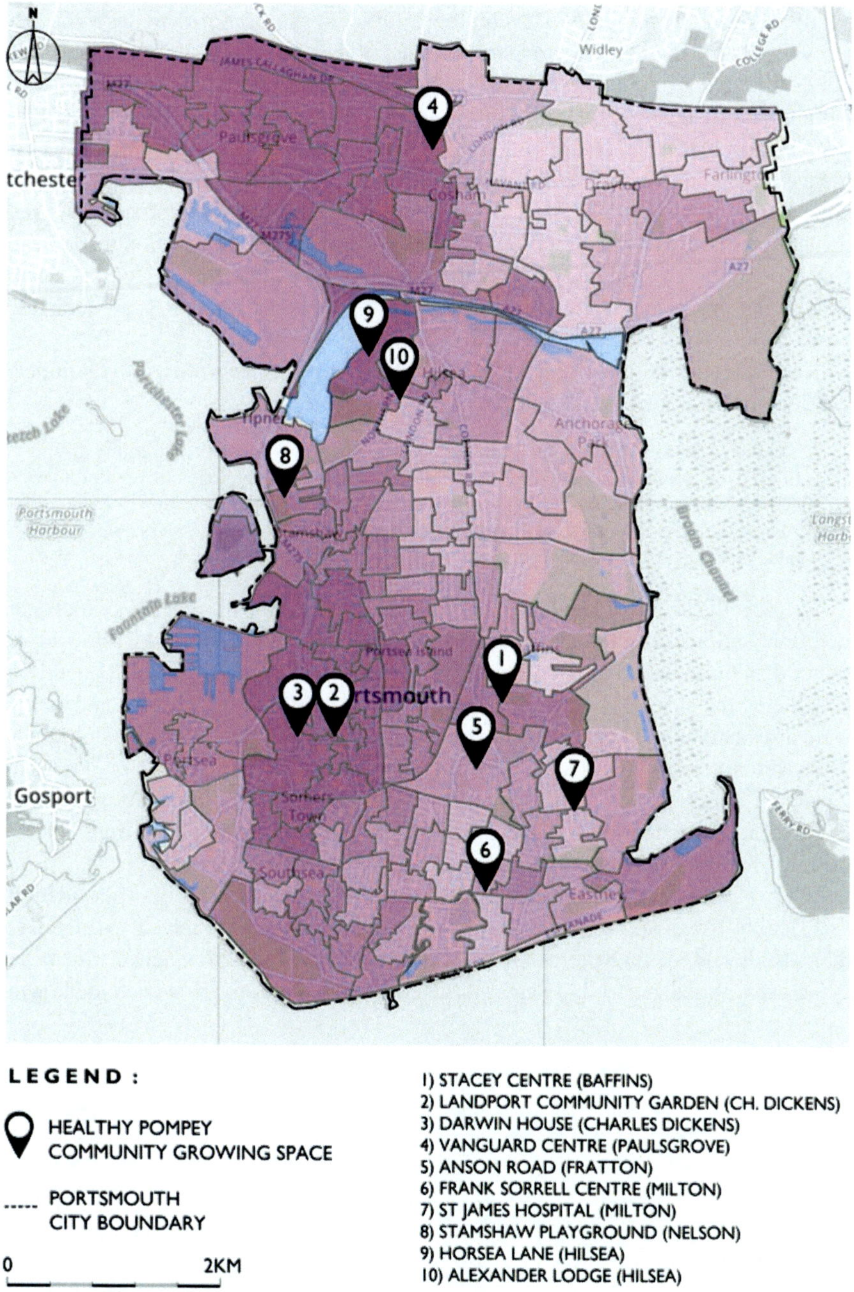

Fig. 11.2 Map of Healthy Pompey community food growing spaces mapped against 2007 areas of multiple deprivation in Portsmouth (*Source* Author with data from Department for Communities and Local Government 2007)
Note Darker shaded areas indicate higher levels of deprivation

Tony Keen, Community Development Officer was assigned to lead the City Growing theme owing to his local experience of developing public health and wellbeing initiatives, including in 2002 setting-up a community allotment project at Milton Allotments, Portsmouth. This project is now well established with over 4,000 people participating since opening; making it one of the best used community allotments in the country and is recognised by the City Council as making an important contribution to improving the health and wellbeing of many of Portsmouth's residents (Portsmouth City Council 2012). Consequently, a key assumption made by the City Council in developing the overarching model was that: 'by increasing access to growing space and training, increased numbers of people will grow their own produce' (Portsmouth City Council 2008b).

Keen was responsible for preparing the food growing aspects of Healthy Pompey's bid document, though points out:

> At the time everyone [City Council] thought: yes, the only way to get people eating healthy is to grow their own food. However, nobody ever asked the people of Portsmouth do you want to grow your own food […] you've got to ask people whether they want to grow […] it was a little bit of a middle-class, top-down: I'm sure all these working-class, poor people would love to grow their own vegetables on their balconies (Int_TK).

As such, the driver of Healthy Pompey's strategy was a willingness to experiment with innovation and pilot area interventions to meet the criteria of Healthy Towns rather than build on existing community development approaches. Consequently, in developing the city's bid it was necessary 'to prove that there was a need and that we were already engaging people. But what they wanted was new ideas, new initiatives. They weren't really interested in anything that we had already done' (Int_TK). This top-down imperative to do something new, to be seen to be innovating, meant a widespread introduction of collective spatial practices in an urban environment that prior to Healthy Pompey only had three community food growing projects.

In May 2009, Healthy Pompey was officially launched, six months after selection as a Healthy Town and this lead-in time presented challenges to the City Council for thematic development. Several new sites for community food growing were identified that corresponded generally to areas targeted for interventions. However, identifying

Fig. 11.3 The Stacey Centre, Portsmouth, and overgrown spaces prior to Healthy Pompey development, circa 2008 (*Source* Photo, left, Author; right, Jones 2019)

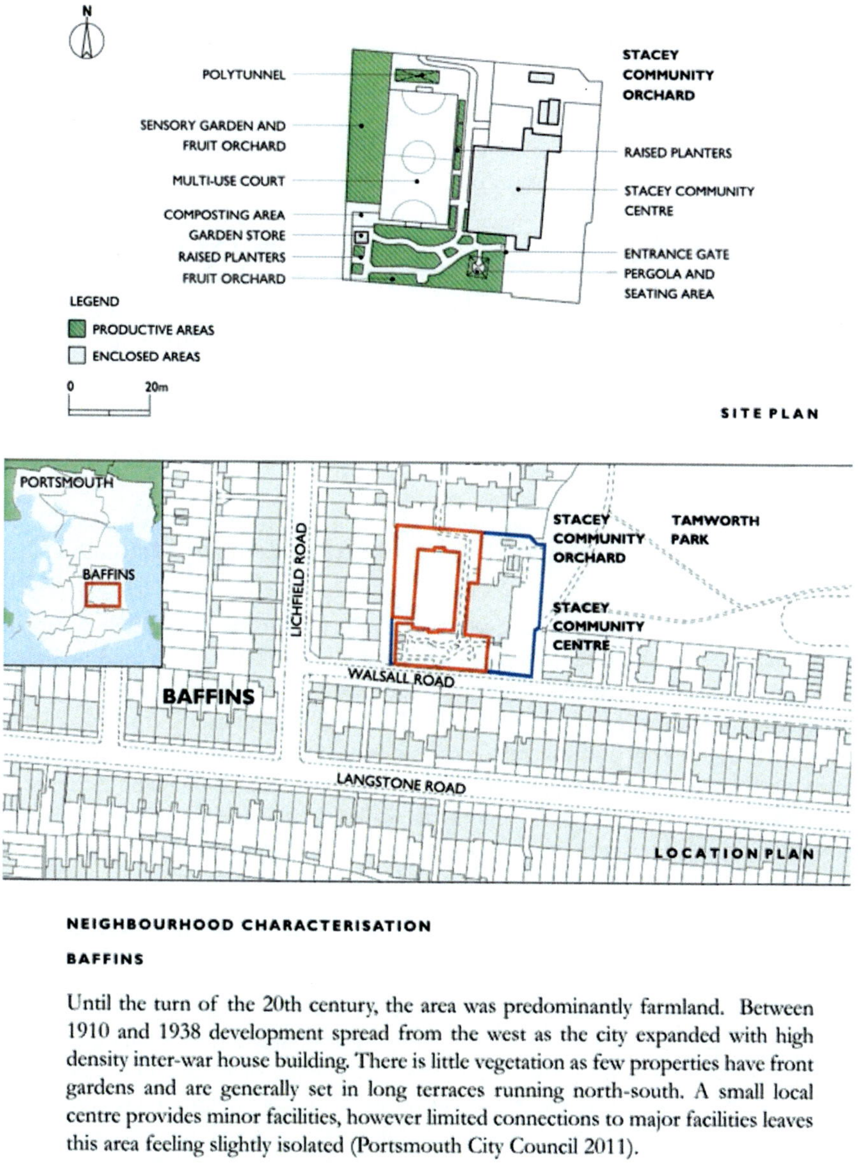

Fig. 11.4 Stacey Orchard site and location plan (*Source* Author)

In the 2015 Index of Multiple Deprivation (IMD) the Portsmouth 014F Lower Layer Super Output Area (LSOA) where the site is located was ranked 9,199 out of 32,844 in England, where 1 was the most deprived and 32,844 the least (Department for Communities and Local Government 2015).

suitable land for growing was difficult due to contamination on large parts of the island created historically by the Ministry of Defence and World War II causing environmental assessments to be required, ruling out several areas. Local knowledge of community facilities also factored in assessing development opportunities with Keen pointing out: 'I'd been a community worker in Portsmouth for 15 years; a good knowledge of all the community centres. We looked all over Portsmouth. Because I've got personal knowledge, I knew that Stacey Centre wanted to develop a garden, a community orchard or something in the space outside' (Int_TK). Aided by this local insight, the Stacey Centre, Baffins was selected as a demonstration space for Healthy Pompey in July 2009. The Centre's locality, popularity with target demographics, and unexploited space also influenced its selection. The community centre had been built by the City Council in the 1960s and later transitioned into a community run facility, but by the late 2000s the external spaces had become neglected and needed attention (Figs. 11.3 and 11.4).

11.2 Development of Social Relations in Systems or Structures

In following the guidelines of Change4Life to build a local coalition to achieve social and urban change, the Healthy Pompey programme worked with several organisations to make it easier for residents to be more active and improve their diet. This included commercial partners, such as ASDA supermarket and local convenience stores to promote fruit and vegetable consumption (Portsmouth City Council 2011). For City Growing, the development of relations between the public and third sector (i.e. non-profit) was particularly significant with relationships developed through a combination of formal structures with local organisations and informal relations with residents.

Contribution of civil society and active citizenship in urban development

Service delivery partnerships were created between Healthy Pompey and the community associations and management committees responsible for selected sites. Several Community Gardeners were recruited to help develop these sites, including Tracey Jones who led Stacey Orchard. The City Council team of Community Gardeners team comprised horticulturalists with experience of developing green spaces and healthy eating projects in the city and in addition to landscape design they worked with the site management and led community activities. Their coordination role was important to programme delivery with Keen pointing out how 'it's really hard to find community gardeners, particularly good ones; the soft skills are really important […] it takes a while to develop, particularly growing food' (Int_TK). An experience of community development and working with people was thus significant to the formulation of grassroots relations, especially to encourage volunteer participation.

The contribution of civil society was a key strategy for Stacey Orchard's development with local people using the community centre having already expressed an interest, as 'embedded resources'. Additional volunteers were specifically recruited, and one-off workdays were held where community groups helped with site activities, whilst on-going volunteering was supported through the links Keen and Jones had to social care and health services referrals (Int_TJ). In winter 2009 work began to clear the site with the support of volunteers. This hands-on approach to urban development was deliberate to support community buy-in, ensuring 'more involvement with the people who are involved in the garden. Because they cleared the garden by hand, the garden belongs to them more. Whereas if we got diggers in, they would've been more detached from that' (Int_TK). A co-design process transpired between Jones, volunteers, and the local community as orchard users, to configure spaces including circulation routes and planting arrangements. Stacey Orchard has fruiting areas and raised vegetable beds, a polytunnel, a composting demonstration area, a cob oven, wild areas, and a sensory garden. Seating areas support either social interaction or peaceful relaxation and plaques with the names of volunteers were placed under fruit trees to promote community ownership of the space and local heritage varieties planted to culturally connect to the area's past (Int_TJ). Information boards on the history of Stacey Orchard additionally provide food growing advice to residents (Fig. 11.5).

Communication between the Community Gardener and volunteers was maintained through informal discussions throughout the development to respond to community ideas. This dialogue was important to developing relations as 'people feel like it is their space and they've got a say in it' says Jones (Int_TJ). Furthermore, communication was significant to neighbourhood development and progressing innovation because 'one of the reasons that people have stayed involved is because they've been consulted' says Keen (Int_TK). The usefulness was demonstrated after Healthy Pompey ended, funding having ceased in March 2011, when a residents' group 'The Friends of The Stacey Orchard' was formed to support ongoing activities and oversee management together with Jones. Since its formation the stakeholders have met weekly to discuss activities and an annual general meeting is held at the Stacey Centre; open to residents to share their views on community events and developments to enable decision-making (Int_TJ). The group also developed relations with several other local groups and food growing initiatives across the city to shares resources, especially volunteers to develop spaces and undertake planting while fund-raising events and community grants (typically £500–£2,000) support continuation of community services.

Collaboration, support, and public communication strategy

Jones also collaborated with experts on the Orchard's design. Speak Out, an accessibility and disability organisation, helped to ensure accessible spaces with relevant footpaths and raised planters. Permaculture experts were consulted to configure the planting and to maximise space; larger trees act as a focal point, while trained 'ballerina type' and patio trees in containers suitable for small spaces were also used. These fruit trees complemented an organic orchard using a permaculture system of agricultural and social design principles while 'edible forest garden' techniques imitated a natural forest with three layers of plants: fruit trees, fruit bushes, and herbs and wildflowers planted by local children. Stacey Orchard needed to act as a Healthy Pompey demonstration garden, especially to showcase urban food growing in small spaces; even if restricted to only a wall or a fence to encourage the replication of

Fig. 11.5 Stacey Orchard's planting phases undertaken by the local community and completed site (*Source* Photo, top row, Jones 2019; other rows, Author)

growing practices among the local population (Int_TJ). In addition, the distribution of home-growing kits to residents sought to build on this approach.

Social marketing was central to the communication strategy of Healthy Pompey which localised and built on the national Change4Life advertising campaign. Relationships were developed with print and broadcast partners as 'communication channels' for the programme (Portsmouth City Council 2011). To promote City Growing, Keen liaised with Programme Manager Hatcher, suggesting:

> [Hatcher] was very good at finding out what we were doing and making sure she knew what we were doing. She was very good at getting other people to link in with what we were doing [...] it was well joined-up the way we worked it out for Healthy Pompey (Int_TK).

Citywide communication was enabled through news coverage aimed at target audiences, especially local families, and demonstrated the relative advantages of learning how to grow fruit and vegetables (see Fig. 11.6). The use of social marketing helped to raise the profile of the campaign whilst promoting its key message: that small changes can lead to large benefits regarding healthy eating and physical activity (Portsmouth City Council 2009b). Brand identity was further reinforced through online and digital communication, personal interaction, and Change4Life signage and Healthy Pompey information boards at cultivation sites (Fig. 11.7).

Since opening, Stacey Orchard has collaborated with local to national level organisations to deliver community services. They partnered with Sustain to deliver Big Dig Edible Gardens Days, a national event to get more people involved in food growing. The Stacey Orchard team work together with Portsmouth Together, a community volunteering scheme organised by the City Council and linked via NESTA (National Endowment for Science, Technology, and the Arts) to an international coalition known as Cities of Service. They connect to the Portsmouth and Southsea Tree Wardens and Tree Council, UK (see Chap. 10 case 'Charles Dickens Orchard Trail', Portsmouth) to jointly deliver horticultural training to residents (Int_TJ). Andy Knight, City Council Arboricultural Officer has participated in these events and provides advice on fruit tree care to Jones and volunteers (Int_AK).

Additionally, Stacey Orchard worked with Portsmouth Friends of the Earth to develop Bee Worlds supporting local biodiversity and participate in walks around local community food growing spaces, whilst featuring in an event held at the University of Portsmouth to bring together local growers to share their knowledge and network. Stacey Orchard donates surplus produce to FoodCycle Portsmouth and works with Abundance Portsmouth, an urban fruit harvesting and distribution initiative. The Orchard also participates in the annual Portsmouth Community Gardens Open Day event, coordinated by Food Portsmouth to promote urban food growing which has led to their partnering on an 'allotment buddy' scheme at Milton Allotments to look after allotment patches for anyone who may be experiencing difficulties and help allocate resources in the city. Gail Baird, Food Portsmouth suggests this collaborative arrangement arose having 'identified a need with some of the people who work there [...] apply for funding independently from the council because of lots of projects the council can't apply for funding for' (Int_GB).

The News

LIFESTYLE REAL LIFE FOOD AND DRINK HOMES & GARDENS CONSUMER GAMING

Project helps people live more healthy lives

FEELING GOOD Debbie Burns with her sons Christopher, six, and 18-month-old Oliver

RACHEL HAWTHORN
Email

08:00 Friday 06 May 2011

Like many parents, Debbie Burns wants her children to know how to lead a healthy life.

So she decided to take them to a Healthy Pompey scheme in the city and see if they could learn fundamental ways to be healthy and live longer.

Every day after school they visited the Stacey Centre in Copnor where the Healthy Pompey team had set up a community orchard for the public to grow fruit and vegetables.

Debbie, 40, along with her six-year-old son Christopher and 18-month-old son Oliver, helped with planting and watering and even took their own Healthy Pompey growing kit

0
comments

HAVE YOUR SAY

Fig. 11.6 Portsmouth newspaper coverage of Healthy Pompey and Stacey Orchard (*Source* Hawthorn 2011)

Fig. 11.7 Change4Life Healthy Pompey signage at the Stacey Orchard (*Source* Author)

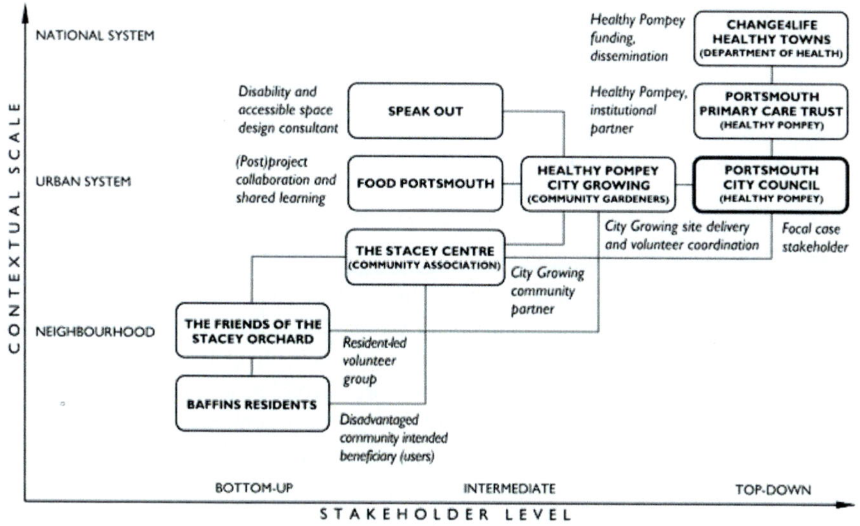

Fig. 11.8 Stakeholders and their relationships in the Stacey Orchard development (*Source* Author) *Note* Focal case stakeholder outlined in bold

Figure 11.8 highlights key stakeholders and relationships in the Stacey Orchard development.

11.3 Provision of Opportunity for Social Empowerment

The provision of opportunities to residents for urban food growing to support behavioural changes through training, workshops, and open days among the local community was evident in this case. City Growing was delivered at several sites, supporting community engagement and according to the City Council, activities were considered to have influenced healthy eating behaviours, either by improved 5-A-Day consumption or maintaining it, and the intention to grow produce in the following year. As such, a 'positive impact' on their state of health was reported by participants (Hencken 2011). Overall City Growing facilitated: 41 volunteers recruited and trained at the various growing sites; 135 community growing workshops held with 1,606 participants; 37 school groups and 597 children participating in workshops; 1,396 people used the training facilities at the Stacey Orchard; 300 people attended Apple Day and Halloween events at Stacey Orchard in October 2010; and 200 people attended Stacey Orchard Open Day in February 2011 (Portsmouth City Council 2011).

Stacey Orchard: a site for public services delivery, community events and activities

The July 2010 launch event for Stacey Orchard was covered in the local news and attended by local councillors and residents while presentations were given on the health benefits of community food growing. Since the launch, public events have been held throughout the year to increase opportunities for community participation and have been attended by several hundred residents (Portsmouth City Council 2011). According to Keen, these events have helped:

> [To] develop ties. This community centre has become more inclusive; between those people, different volunteers, different ages, backgrounds […] community aspect of people meeting and mixing with people they might not have mixed with so much before and just amongst themselves they had that social contact (Int_TK).

An Easter Fun Day is arranged for children to take part in vegetable sowing and environmental activities and a Summer Fayre is held for community fruit harvesting while children hunt for 'mini-beasts' as part of the Woodland Trust's Nature Detectives programme. A Bee Friendly fun day is organised in the summer and a crafts school involving gardening activities is held during school holidays. A garden party celebrated the 50th anniversary of the Stacey Centre in August 2018 and at Halloween there is pumpkin carving and winter vegetable sowing (Fig. 11.9).

At Stacey Orchard training is provided to volunteers and residents; typical sessions involving guidance on fruit tree cultivation and pruning led by the Community Gardener with further information provided at open days and to visitors on an ad-hoc basis (Int_TJ). Following the wishes of volunteers, a grant from the Earth Pathways Co-operative enabled several willow workshops which taught volunteers how to make items out of natural willow, including vertical food growing supports and bird feeders, these skills are then passed on to local people who attend the open days. Additionally, gardening activities are run by the Stacey Friday Club for adults with physical and learning disabilities to widen the support of community development.

Healthy Pompey: getting the message across early to local people to promote healthy lifestyles

Learning activities at Stacey Orchard are especially targeted towards children in the community and delivered with local partners. Primary schools use the space for outdoor education with pupils aged 4–11 taking part in activities such as seed sowing, herb smelling, and fruit and vegetable identification. The Baffins Nature Nursery based on the principles of the Forest Schools movement also use the space for education where young children learn about cultivation and harvesting, and organic produce is used in snacks to promote a healthier diet (Fig. 11.9). Local teenagers are additionally involved at Stacey Orchard, including one volunteer who went on to obtain an allotment due to developing an interest in food growing (Int_TJ).

As such, working with families and young children was a key approach to Healthy Pompey's aim to instigate local inhabitants' behaviour change with City Growing interventions. In this regard, Gardner et al. (2012) argue that habit formation occurs early in life and efforts to influence behaviour are most effective at this critical stage to form healthy habits to take into adulthood. In supporting education locally, Keen suggests:

> It's got to be a holistic approach. The education we are giving to those kids: they might be young carers or from disadvantaged homes, so it is a way of justifying it, building in that

Fig. 11.9 Stacey Orchard launch event in July 2010, community open days, and children's activities (*Source* Jones 2019)

> sustainability […] as they grow-up they know it's here. Knowing that space is there, and it means their families too because they tell their parents; the younger you can get the better (Int_TK).

By increasing awareness and user involvement, capacity building for behaviour change amongst residents was developed through City Growing and thus helped to ensure spatial practices were supported through cultural shifts in the city.

11.4 Reflection of Socio-spatial Development Practice

The impact of urban interventions in food growing was monitored, analysed and evaluated using feedback systems to capture changes regarding healthy eating and physical activity (Portsmouth City Council 2009a). However, demonstrating change was challenging, as understood by a 2007 evaluation of an earlier obesity programme in Portsmouth:

> The majority of healthy eating projects can provide qualitative but not quantitative data, and there is no hard evidence to illustrate that any of these projects are having a direct impact in terms of weight loss. It is a similar pattern for physical activity, where a number of individual projects have successfully met their aims/ objectives, but evidence of behaviour change is mostly anecdotal (Brilliant Futures 2010, p. 6)

In order to measure the impact of Healthy Pompey, local health consultancy Bodymorph was commissioned to evaluate the effectiveness of each theme using a range of methods. One element involved a two-year measuring programme to assess participants health; monitoring change across a variety of biometrics, quality of life and lifestyle measures (Portsmouth City Council 2011). However, in appraising City Growing's social impact this created conflict as Keen points out:

> [Because] their whole focus was on weighing people and physically assessing them. They wanted to measure all our volunteers and at the end of it measure them again […] The assessors said they didn't think we enabled them to do it enough […] as soon as someone says: we want to assess you, we want to focus on your weight; it makes you feel worse […] making people feel bad before they even started. If people's wellbeing is low, then they're never going to improve their physical health (Int_TK).

Consequently, other baseline data and evaluation measures were used to assess this theme, with information being obtained via questionnaires; one focussed on food growing sites and the other on the impact of the growing kits that had been distributed. Through collecting participant data from theme leads, Bodymorph examined user engagement, demographics, the relationship between involvement in growing activities and healthy behaviour, and the indirect impact on health behaviour (Hencken 2011). Across the City Growing sites, of the 150 participants who took part in the evaluation, 69% intended to cook what they had grown while regarding quality of life and positive health improvement, 38% reported improved fitness and 31% noted increased self-worth (Portsmouth City Council 2011).

Outcome measures for City Growing's evaluation were primarily quantitative, specifically: participant numbers and numbers of new food growers and growing groups; new growing sites; volunteers recruited; home growing packs distributed; information packs distributed; and training sessions held (Portsmouth City Council 2009a). Weighing produce to assess the amount grown was also initially undertaken though this data was not included within the final evaluation, likely on the basis it did not appear considerable (Int_TK). Healthy Pompey led to the development of urban green infrastructure in areas of higher health inequalities regarding the maintenance of healthy weight, with 10 new sites being developed in economically deprived neighbourhoods, thus helping to achieve an aim of Healthy Pompey. For Portsmouth, it suggests the programme has enabled more equitable access to urban material resources and contributed towards the idea of 'spatial justice' as called for by Soja (2010).

The City Growing success criteria sought an 'increase in people benefiting from direct access to nature' (Portsmouth City Council 2009a). Contextually, the provision of local opportunities to improve wellbeing are important. Keen says: 'there are still children who live in Paulsgrove who've never been to the beach and there are kids who live in Landport and Somerstown who've never gone to the countryside' (Int_TK). As such, City Growing created openings for disadvantaged people to access public green space and participate in community food growing with 6,285 people visiting sites between 2009 to 2011 and an additional 2,448 Home Growing Kits were further distributed to residents (Portsmouth City Council 2011). Yet, the widespread distribution of growing kits also highlights a basic approach to measuring 'impact', with Keen stressing:

> The reason we did it was part of the outcomes that the funders wanted was substantial numbers. So, if we send out 10,000 packs that's 10,000 people who are growing [...] it is a narrow way of assessing this, the success of something. Whereas a lot of assessment of community work is not necessarily quantifiable, it's anecdotal (Int_TK).

This insight reiterates the evaluation of earlier obesity interventions in Portsmouth. It suggests that assessing the aim of the Change4Life campaign for long-term behaviour change in diet and activity is challenging. Obstacles here included the short-term focus on quantitative assessment and the requirement to be seen to be delivering effective interventions. Furthermore, the situation found in Portsmouth is supported by feedback across the coalition of aspiring healthy towns; the evaluation of impact being considered difficult to report by stakeholders within the three-year programme (Cummins et al. 2016).

The progression of social innovation, imitation, and knowledge sharing

Sustainability was positioned as key to the Healthy Pompey programme to ensure that residents continue to gain from the investment in their health beyond the initial funding period (Portsmouth City Council 2011). The continuation of food growing practices locally is however challenged, especially as three of the growing sites later closed due to the day centres where they were located being closed down (Food Matters 2015). For example, the former Vanguard Centre, Paulsgrove was a well-used disability centre run by the City Council that was closed within six months of developing an accessible garden in 2010 at a cost of £45,000, the site having since

Fig. 11.10 Healthy Pompey community garden at Vanguard Centre, Cosham (prior to closure) and the Community Garden at Treadgolds Museum, Portsmouth (*Source* Photo, left, Portsmouth City Council 2011; right, Author)

been redeveloped and the garden removed (Fig. 11.10). The issue is attributed to local political changes in Councillors and subsequent revisions to strategic priorities of the City Council (Int_TK). As such, earlier consultation work is proposed between urban stakeholders and end users in identifying initiatives, especially area interventions to seek greater certainty in long-term operation. The closure of centres also highlights a further issue in developing spaces in community centres now managed by community associations (i.e. dissolved from City Council) in that they may close due to on-going funding challenges, including Stacey Centre.

The funding of Healthy Pompey was significant in creating the opportunity to develop a strategic level citywide approach to social innovation. Conversely, the short-term focus of the programme restricted the progression of innovation beyond this funding period, with Keen explaining:

> More money in three years than I've ever had in 30 years [...] there is all this money, then suddenly there was no money. Whereas they could've saved half the money from Healthy Pompey and eked it out over the next 10 years that would have really benefitted the longevity of the project [...] but if we hadn't had that substantial amount of money, we wouldn't have been able to do it (Int_TK).

Furthermore, the funding of City Growing's Community Gardeners to coordinate community activities was only intended to be funded for the duration of Healthy Pompey, threatening their continuation (Portsmouth City Council 2009a). This meant a justification had to be made to the City Council to retain their participation across sites, albeit now as a part-time role, and was supported by demonstrating the number of participants still engaging with the spaces developed (Int_TJ).

The benefit of retaining the Community Gardeners was demonstrated in supporting the replication of cultivation spaces in Portsmouth. For example, they supported development of sites within economically deprived areas including Wymering Manor, Cosham in 2014 and at the Somerstown Community Garden (Omega Centre) and Somerstown Adventure Playground in 2015. They also helped Food Portsmouth in 2017 to develop the Treadgolds Museum Community Garden with Jones coordinating their volunteer days (Fig. 11.10). As in the case of Stacey

Orchard this is also an accessible garden, and the Community Gardeners are helping to develop accessible gardens for social housing residents living with dementia. Consequently, the City Growing lead has been approached by other local authorities to share development knowledge of community food growing (Int_TK). Yet, in terms of Healthy Pompey influencing policy in Portsmouth, the programme is suggested by Keen as:

> [Having] come and gone, been forgotten [City Council] have dumped social marketing now, it was a bit of a luxury, popular at the time [...] you feedback to the funders Central Government. This is the trouble. As soon as the last Labour Government left, the whole ethos of the Council changed (Int_TK).

Nationally, this situation was compounded by the Healthy Towns programme spanning a change in government which likewise led to a shift in strategic priorities and funding (Cummins et al. 2016). Locally, there is no reference to the Healthy Pompey in Portsmouth's Joint Strategic Needs Assessment (2013). While Change4Life branding can still be found in other cities, the Portsmouth brand has been discontinued by the City Council and the social marketing advertising, website and social media channels no longer exist, thus affecting its intended legacy.

To share learning nationally with other Healthy Towns, Portsmouth did participate in a Healthy Towns Conference in 2011 and further dissemination of Healthy Pompey was undertaken through participating in the Healthy Towns evaluation (Hencken 2011). Quarterly monitoring reports were made over three years on thematic interventions and the Healthy Towns final report of the Department of Health was published in 2016. Among its key findings were that participating 'towns felt under some pressure from a variety of stakeholders to demonstrate positive results' and consequently this influenced approaches to innovation and delivery of interventions, causing an 'innovation and evidence paradox' (Cummins et al. 2016, p. 4). This finding was shown to be accurate for Portsmouth because of the focus on achievable short-term outcomes (e.g. numbers of interventions and participants involved). Consequently, based on the execution of the national programme, to optimise future delivery of similar systems-based interventions, the report made key recommendations, including: the requirement for greater direction; to enhance and better develop a practice-focused evidence base; and the development of approaches that help foster innovation locally (Cummins et al. 2016).

11.5 Case Discussion and Conclusions

In this section I discuss the interactions between key social innovation processes and summarise my findings. I conclude the following section with a reinterpretation of this case in key processes and spatial scales of analysis.

In summary, I analysed how a local authority sought to develop urban food growing within a systems-based health programme. The City Council sought to address an identified need to tackle obesity levels and urban health inequalities through policy-based social innovation. In so doing, they developed and built on relationships with community organisations within the urban system to form service delivery partnerships, whilst activating citizens to help deliver productive green infrastructures. The provision of spaces for community food growing citywide contributed to 'distributive justice' in the socially just allocation of resources to support residents' behavioural change, including opportunities for skills development, community participation, and public access to space. Sharing of innovation learning took place directly between the City Council and Central Government through evaluations to share knowledge of systems-based interventions.

The identification of societal challenges at a national level was significant here to prompting social innovation. The local solution proposed by Healthy Pompey sought to improve urban infrastructure and facilitate services in order to promote healthy lifestyles. A key strategy was the introduction of several public cultivation sites in the city, a city which had limited existing social activity around food growing. As such, this approach to innovation is suggested as the 'recontextualisation' of a social practice or behaviour. However, there was limited evidence of identification activity for these interventions in their conception at the strategic level (aside from local knowledge of the City Growing lead of the programme) to understand the effectiveness of the proposed solution, beyond identification of disadvantaged communities most at risk of obesity. As such, the needs of the urban population and approach to satisfying those needs was determined by the City Council.

The planning of an innovative whole town approach and partnership between the local authority and the community suggested tensions and contradictions in the need to be seen by its funders to be experimenting with novel spatial practices, though regarding local needs identification and area interventions in general, community-run spaces in Portsmouth were supportive. This process was shown in the City Growing theme of Healthy Pompey, including the development and continuation of Stacey Orchard, however the potential to identify and satisfy end-user needs across the programme might have been greater if intervention design and delivery had been better aligned. Therefore, I suggest that future innovation strategy should ideally build in the necessary time and commitment for robust identification processes, including an assessment of which service user needs are deemed important for policy and practice prior to implementation.

On analysing how social relations were developed, the case indicates that forming effective partnerships for public service delivery was significant to advancing innovation. In order to develop community good growing spaces the City Council needed to form links with local community associations to reach their target audience, with local organisations ideally being representative of the residents. Generating an 'active citizenship' in local volunteers helped to transform sites, though more importantly enabled a process of community buy-in from residents as stakeholders to advance social innovation. This was enabled through a development sequence of gaining access to a neighbourhood, introducing the idea for an intervention, and engaging communities in the intervention. Here an understanding of neighbourhood context in order to provide a service used by residents was necessary and an appropriate site selected so that beneficiaries would value the transformation of neighbourhood space. However, in proposing strategic level ideas, I suggest that resident input in this case should ideally have supported their initial forming, including how local needs might have been addressed to assist project development.

Significant to the case's evolution, was the creation of opportunities for the community to shape ongoing neighbourhood development through participatory decision-making. Opportunities were given to residents to be involved in decisions through open meetings throughout the development, helping to ensure a sense of 'co-ownership' over cultivation spaces. This facilitation was demonstrated in collaborations between City Council representatives (as professionals) and local volunteers (as users) to shape spaces and services after their initial planning. However, to understand any governance changes to relations within the urban system, this process is implied to centre on the development of informal relationships between the local authority and the residents rather than implementing any formal and lasting decision-making structures.

The ethos of the Healthy Towns programme and Change4Life social marketing campaign focussed on developing partnerships between local authorities and cross-sector stakeholders to support urban change. While the community was given a voice to 'have a say'; the fundamental decisions were shown to be made by the City Council, especially during the commissioning and planning phases of spatial interventions. It would appear therefore that this case raises wider issues around the equality of state-civil society partnerships and ability to share power in collaborative city shaping.

Regarding the provision of opportunity, my findings indicate that a two-stage strategy of 'engage and enable' was central to implementing policy-based innovation. This top-down approach focused on communicating changes to affect individual behaviour and stimulate urban inhabitants to behave differently through 'nudge-style' interventions. Social marketing was employed here via a mass media campaign to create awareness amongst target audiences and supported later through interpersonal channels during provisional activities to encourage innovation adoption. Healthy Pompey's interventional approach sought to affect individual's choices as 'consumers' through informing and educating. This process was enabled by providing opportunities to residents to enhance capabilities through the development of 'self-efficacy', i.e. the ability and know-how of people to implement and maintain new behaviour (see e.g. Bandura 2001), as a key aim of the Healthy Towns programme in order to change people's behaviour to live healthier lives.

The provision of Growing Workshops played a key role to empower communities to develop new behaviours to support healthier lifestyles and in-part helps to explain why community food growing continued at several city spaces after the programme. This suggests an emerging compatibility of spatial practice to context (e.g. Portsmouth) in meeting the cultural values, experiences, and needs of local inhabitants as innovation adopters. In this sense, Rogers (1995) argues that compatibility with innovation helps the individual give meaning to the new practice so that it is held as familiar. My observation therefore indicates a need to further understand the interaction of residents' spatial practices and their material contexts to innovation adoption. Such an understanding would advance an intervention strategy in behaviours towards greater reflection on why social and spatial practices are formed and repeated, especially in seeking to influence the habit formation of future generations of residents through social innovations.

The reflection of socio-spatial development practice was shown to be influenced by the need to demonstrate impact, e.g. to make the case for continued Central Government funding, rather than a directive to learn from local experience or to base future innovation approaches in the city. As such, data gathering, and the use of feedback systems was shown to be focussed towards evaluating the number of interventions and participation of Portsmouth's residents in determining whether the local programme had met its objectives. Through an examination of the success criteria of Healthy Pompey, it is found that urban infrastructure improvements are evident and that opportunities were provided to empower residents to take responsibility for their health. This process is suggested to contribute to the potential reduction of urban health inequalities because interventions were generally delivered equitably as evidenced in the targeting of resource-poor areas and communities with highest population need. However, in evaluating the criteria of whether the health of Portsmouth's residents has been 'sustainably improved' through changes in lifestyles and behaviours without rigorous assessment is difficult. In particular, the tracking of changes and overall outcomes with, for example, a before-and-after study of target populations proved challenging (despite the attempts to measure body mass index).

Lastly, I suggest a cultural shift can be observed across Portsmouth in the reproduction and growth of spatial practices through the development of community food growing spaces from the public sector and civil society following Healthy Pompey (see Chap. 10 case 'Charles Dickens Orchard', Portsmouth). However, evidence of shared learning of the specific programme within the City Council or amongst other local stakeholders appears to have been limited owing to its focus being directed toward the national level health programme.

11.6 Interpretation of Social Innovation Pattern

The visual narrative map reveals the large amount of identification actions at the national scale in the opportunity to trial systems-based interventions in order to support aspiring 'Healthy Towns', including Portsmouth (Figs. 11.11 and 11.12). This Central Government strategy causes identification processes within the urban system to target area interventions. The development of relationships then follows from the urban to neighbourhood scales to connect Portsmouth's community centres as delivery partners and to activate local people as volunteers. After 'Healthy Pompey' launches, provision activity occurs citywide to help residents, especially those within deprived areas, to develop 'self-efficacy' in their confidence and ability to adopt new behaviours to support healthier lifestyles. This provision continues within the Baffins neighbourhood after Healthy Pompey ends through the continuation of activities at Stacey Orchard. Reflection processes by the City Council to support internal learning and local development is implied to be limited with monitoring, evaluation, and dissemination being directed toward the Healthy Towns programme and Central Government in order to demonstrate impact.

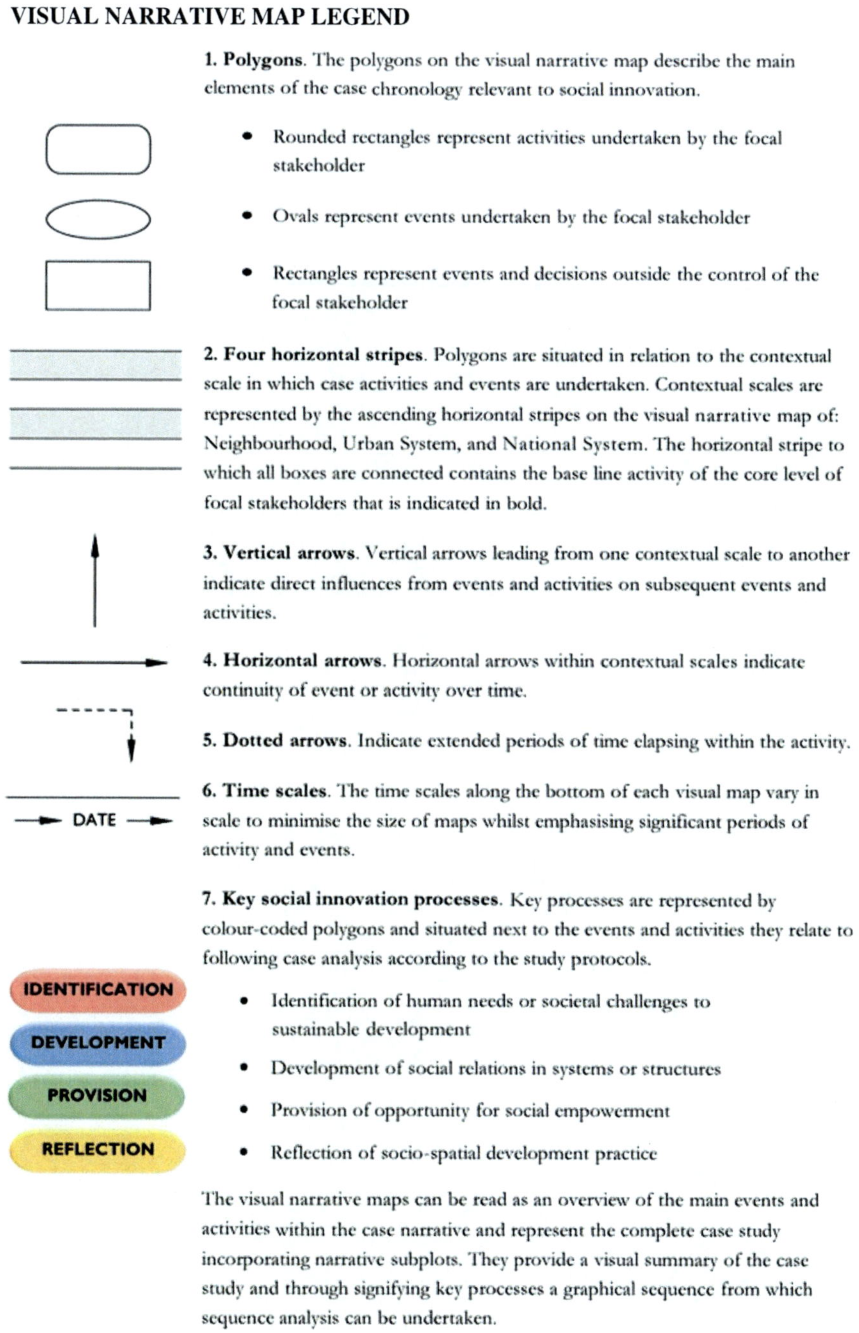

(*Source* Author)

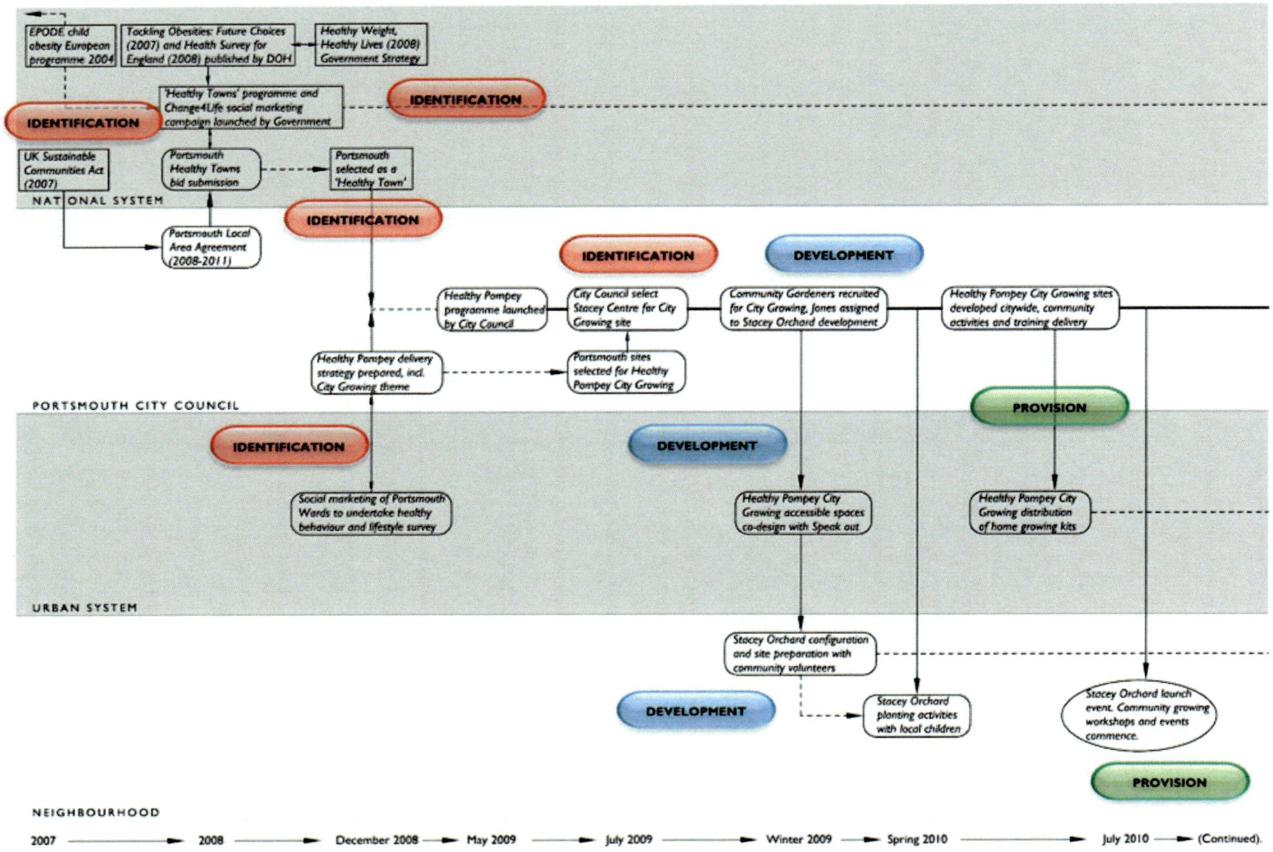

Fig. 11.11 Visual narrative map of the Stacey Orchard development with key social innovation processes indicated across scales. Map sheet 1 of 2 (*Source* Author)

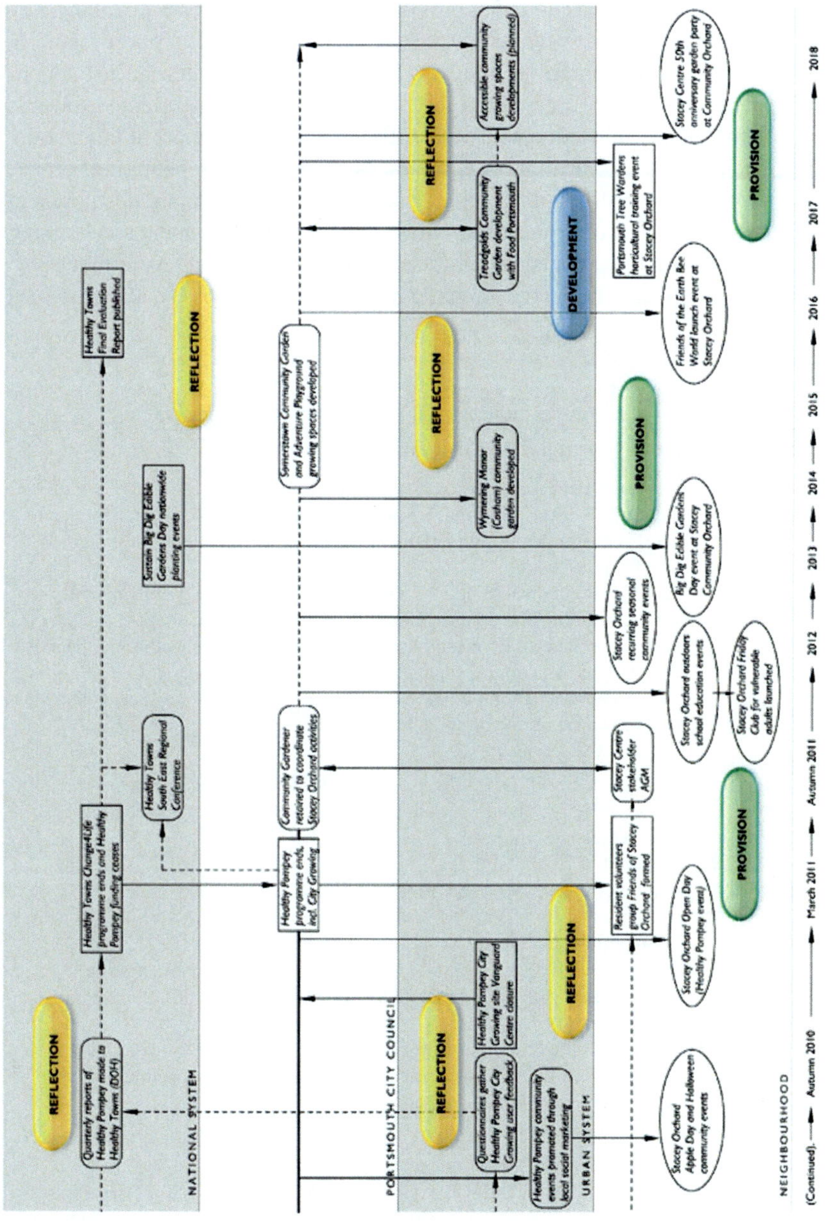

Fig. 11.12 Visual narrative map of the Stacey Orchard development with key social innovation processes indicated across scales. Map sheet 2 of 2 (*Source* Author)

Regarding reflection processes, the case reveals similarities to the 'Racehill Community Orchard', Brighton & Hove (see Chap. 7) in its small deviation from the proposed conceptual framework with monitoring and evaluation by urban stakeholders occurring both during and after provision activity. This is partly explained by similarities between these two cases in the need of stakeholders to report regularly to urban development programme funders. However this case needed to demonstrate impact in relation to specified objectives rather than being oriented towards sharing of innovation knowledge with other cities to widen the impact of innovation. Locally, the ability to advance innovation here by avoiding the repetition of errors, promote successful initiatives, and make recommendations for future development programmes was impacted. Especially, owing to innovation assessment and learning outcomes by passing the urban system in feeding back to national level funders.

Table 11.1 provides a summary of contextual influences across spatial scales on development.

Table 11.1 Summation of contextual influences on Stacey Orchard development

Contextual scales	Influences on project development
National system (Macro scale)	• Tackling Obesities: Future Choices (2007) and Health Survey for England (2008) Government reports, rising trend in UK obesity levels, highlighted need for interventions • Healthy weight, healthy lives: a cross-government strategy for England (2008), 'Healthy Towns' programme (2008–2011) and Change4Life, set targets to reduce child obesity levels in urban environments through healthy lifestyle and behavioural changes • UK Sustainable Communities Act (2007), set guidelines for local authorities to reverse urban trend of community decline (in ongoing loss of local facilities and services)
Urban system (Meso scale)	• Portsmouth Local Area Agreement (2008–11), high obesity rates, especially people living inactive lives in deprived wards, demonstrated need for innovative health initiatives • Portsmouth 'Healthy Pompey' bid submission, key theme City Growing within bid strategy to create food growing spaces in city • Social marketing survey and health insight of family lifestyles, identification of territorial health initiatives; • Co-design with Speak Out, facilitated accessible space configuration • Collaboration with Food Portsmouth, supported Treadgolds Community Garden; • Collaboration with Portsmouth Friends of the Earth and Portsmouth and Southsea Tree Wardens facilitated horticultural training, community engagement, and mutual support
Neighbourhood (Micro scale)	• Area of high obesity levels and inactive lifestyles amongst target audience, created need for interventions • Local community centre, unexploited public land identified for development • Interested resident volunteers and resident-led The Friends of Stacey Orchard group, devolves spatial governance and supports long-term development

Source Author

References

Bandura A (2001) Social cognitive theory: an agentic perspective. Annu Rev Psychol 52(1):1–26
Borys J-M, Valdeyron L, Levy E et al (2010) EPODE—a model for reducing the incidence of obesity and weight-related comorbidities. Eur Endocrinol 9(2):116
Brilliant Futures (2010) Healthy Pompey: eat well, move more, live longer. Portsmouth City Council, Portsmouth
Carvel J (2008) "Healthy towns" to fight obesity. In: The Guardian. https://www.theguardian.com/society/2008/nov/10/obesity-healthy-towns1. Accessed 1 Jul 2018
Cummins S, Ogilvie D, White M et al (2016) National Evaluation of the Healthy Communities Challenge Fund: The Healthy Towns Programme in England. Final report to the Department of Health. LSHTM, London
Department for Communities and Local Government (2007) Indices of multiple deprivation 2007. Department for Communities and Local Government, London
Department for Communities and Local Government (2015) Indices of deprivation 2015: index of multiple deprivation and domain indices. Department for Communities and Local Government, London
Department of Health (2008a) Health survey for England 2008. HM Government, London
Department of Health (2008b) Healthy weight, healthy lives: a cross-government strategy for England. HM Government, London
Department of Health (2010) Change4Life one year on. HM Government, London
Food Matters (2015) A review of the food system in Portsmouth. Food Matters, Brighton & Hove
French J, Blair-Stevens C (2006) It's our health! Realising the potential of effective social marketing. National Social Marketing Centre, London
Gardner B, Lally P, Wardle J (2012) Making health habitual: the psychology of "habit-formation" and general practice. Br J Gen Pract 62(605):664–666
Government Office for Science (2007) Tackling obesities: future choices-project report Government Office for Science. HM Government, London
Hawthorn R (2011) Project helps people live more healthy lives. In: News. www.portsmouth.co.uk/lifestyle/real-life/project-helps-people-live-more-healthy-lives. Accessed 17 Aug 2016
Hencken C (2011) Healthy Pompey programme evaluation: successes and challenges. Bodymorph, Portsmouth
Jones T (2019) The Stacey Community Centre. https://www.facebook.com/pg/TheStaceyCommunityCentre/photos/?tab=albums. Accessed 1 Sep 2019
Portsmouth City Council (2008a) Portsmouth Local Area Agreement 2008a-11. Portsmouth City Council, Portsmouth
Portsmouth City Council (2008b) Portsmouth City Council—overarching programme logic model. Portsmouth City Council, Portsmouth
Portsmouth City Council (2009a) Urban growing. Portsmouth City Council, Portsmouth
Portsmouth City Council (2009b) Healthy town: Portsmouth. Portsmouth City Council, Portsmouth
Portsmouth City Council (2010) Healthy city standing scrutiny panel. Portsmouth City Council, Portsmouth
Portsmouth City Council (2011) Healthy Pompey impact report: paving the way for a healthier future. Portsmouth City Council, Portsmouth
Portsmouth City Council (2012) Ten years of the Community Allotment at Milton Piece, Portsmouth 2002–2012. Portsmouth City Council, Portsmouth
Portsmouth Primary Care Trust (2008) Healthy weight strategy for Portsmouth City: 2008–2011. Portsmouth City Council, Portsmouth
Rogers EM (1995) Diffusions of innovations, 4th edn. The Free Press, New York
Soja E (2010) Seeking spatial justice. University of Minnesota Press, Minneapolis

Part IV
Social Innovation Processes, Patterns and Contextual Forces in Urban Food Spaces: A Cross-Case Analysis

Chapter 12
Social Innovation Processes and Patterns in the Production of Urban Food Spaces

Abstract This chapter draws together observations from across the cases studies to address the research question: 'Are there any observable patterns to the key social innovation processes undertaken in the production of urban food spaces and, if present, what accounts for these?' In so doing, it examines the relationships between social innovation processes to establish a common pattern of innovation through which urban stakeholders collaboratively produce spaces for food growing in cities. The chapter builds on the proposed conceptual framework key processes of: (1) Identification of human needs or societal challenges to sustainable development; (2) Development of social relations in systems or structures; (3) Provision of opportunity for social empowerment; (4) Reflection of socio-spatial development practice. The identified pattern features events and outcomes which are dependent on both space and time as important components associated with key innovation processes to support this sequence. Movement back and forth between innovation processes is found to be important to progress social innovation in urban development, while changing environmental conditions suggest that key processes are potentially adaptable depending upon circumstance.

Keywords Social innovation processes · Innovation patterns · Urban food growing · Co-production of space · Social innovation · Process research

12.1 Patterns to Social Innovation Processes in Spatial Production

In order to enable the identification of innovation patterns in the production of food growing spaces by urban stakeholders, I used process research approaches (Chap. 2). In my study, patterns are interpreted as the sequences found in common and observed across space and time. These innovation patterns or 'motors' as characterised by Poole et al. (2000) help to explain how and why changes unfold in developing process models. Furthermore, along with narrative explanation, the recognition of progressive structures helps to 'describe the form and content of developmental and change theories' (p. 27). In the following, I advance the conceptual framework proposed in Chap. 4 by recognising and accounting for patterns of key social innovation processes.

N. Ardill, *Growing Food in Cities*, Cities and Nature,
https://doi.org/10.1007/978-3-030-98475-5_12

A common pattern to social innovation in the production of urban food spaces

All city case studies, both within the UK and across Europe, are exemplified by the provision of opportunities for social empowerment; their main intent being the production of spaces for community food growing to satisfy local needs. I will now summarise the in-depth analysis of six cases within two UK cities of Brighton & Hove and Portsmouth.

In Brighton & Hove, the residents' co-operative at The Bevy Edible Garden produced a community hub with interconnected green spaces as a social enterprise; Racehill Community Orchard was produced by a food advocacy organisation to demonstrate urban food growing practices within a citywide development programme; and The Keep Community Orchard was produced by a local authority to experiment with planning approaches to sustainable development. In Portsmouth, the residents' co-operative at the Southsea Greenhouse produced a neighbourhood facility and social enterprise to support community development; The Charles Dickens Orchard Trail was produced by an environmental organisation to develop a linear fruit orchard as a 'productive urban landscape' for local people; and lastly, Stacey Community Orchard was produced by a local authority to promote food growing activities as part of a national programme to tackle obesity.

In each of the above cases, a substantial identification of needs and development of social relations activity was needed to create opportunities to satisfy needs, e.g. local insight had to be gathered and interpreted, stakeholders mobilised, and service delivery partnerships formed. This indicates that as a foundation to provide opportunities, urban stakeholders must assume prior identification and development activities. Building on the account of the case studies, a basic pattern on how innovation develops can be observed (Table 12.1). Each organisational level of urban stakeholder navigated through a basic sequence of key social innovation processes, as follows:

After the identification of needs to assess and frame contextual openings and constraints, the next process involves generating support and coordination through developing cross-sector relations to obtain necessary resources to make initiatives happen. In cases where developing relations was ineffective (e.g. Southsea Greenhouse, Portsmouth), the initiatives' strategic brief and development plan were redesigned, including the aim of the initiative, its specification, and people and spaces involved to gather additional backing or means. In cases where developing relations was shown to be effective, stakeholders generally progressed to the provision of opportunity through developing a pilot initiative (e.g. Racehill Community Orchard, Brighton & Hove). Common to all cases is the provision stage when urban initiatives were delivered and the opportunity to satisfy needs is brought about. If provision activity was not achievable or did not last, the initiative re-brokered access to

Table 12.1 Basic pattern of key social innovation processes in the six UK-based cases

Case	Event (Inputs and trigger for action)	Identification of human needs or societal challenges to sustainable development	Development of social relations in systems or structures	Provision of opportunity for social empowerment	Outcomes (Observable changes and indicators)	Reflection of socio-spatial development practice
The Bevy Edible Garden, Brighton and Hove, UK	Prospective loss of neighbourhood asset amid the on-going closure of neighbourhood social spaces	Formation of a campaign group, consultation of residents and preparation of development brief for social enterprise	Created intermediation spaces; Secured support for community share issue and grant application	Increased community participation and self-organisation; Developed community resource	Bevendean Community Cooperative (800+ members); Community hub used annually by 70 groups	Shared learning; Local imitation of resident-led development and community business model
Racehill Community Orchard (Harvest), Brighton and Hove, UK	Unsustainable national food system; Opportunity to increase urban food growing and the number of participants	Deliver an urban growing programme via Local Food Fund beacon project grant	Formed cross-sector partnerships; Coordination and connecting	Developed personal, material, and cultural capabilities within the urban system	Exemplar food growing space; 54 food growing spaces (1.19 ha); 65 training courses (550+ attendees); Community food growing 4,000 volunteers per year	Dissemination activities at multiple scales (urban to national); Formation of UK Sustainable Food Cities Network
The Keep Community Orchard (PAN06), Brighton and Hove, UK	Change in local authority leadership and strategic direction	Experiment with integrating food growing within urban planning and sustainable development via a prototype development	Collective and insider lobbying; Participatory planning; State-civil society partnership	Vertical scaling-up of development approach into planning policy	Adoption of planning policy; 44 applications 712 m^2 food growing space within proposals (over 1 year); Sustainable urban development framework	Learning with policy-based innovation shared to other localities

(continued)

Table 12.1 (continued)

Case	Event (Inputs and trigger for action)	Identification of human needs or societal challenges to sustainable development	Development of social relations in systems or structures	Provision of opportunity for social empowerment	Outcomes (Observable changes and indicators)	Reflection of socio-spatial development practice
Southsea Greenhouse, Portsmouth, UK	Urban decline and lack of seafront facilities	Create a neighbourhood facility to support community development via social enterprise	Brokered cross-sector support to access material resources	Increased community participation; Delivered skills workshops	Southsea Greenhouse Community Cooperative (200+ members); 180 skills workshops (300+ participants)	Cyclical sequence of prototyping and adaption of initiative
The Charles Dickens Orchard Trail, Portsmouth, UK	Potential loss of neighbourhood open space and lack of utility to residents	Encourage local foraging and food growing; Connect neighbourhoods and provide a free shared resource	Facilitation by City Council Officers; Co-design in partnership with professionals	Brought communities together to transform the urban environment	165 fruit trees planted by 250 resident participants across three planting phases	Learning and adoption of innovation across local neighbourhoods
Stacey Community Orchard (Healthy Pompey), Portsmouth, UK	Obesity levels and inactive lifestyles of local families; Opportunity to increase urban food growing to improve health	Trial systems-based interventions to modify urban environments via Government funding	Partnerships with community associations for service delivery	Development of self-efficacy in the ability and know-how to implement and maintain new behaviours	Demonstration urban orchard; 10 food growing spaces; 135 food growing workshops (1,606 participants); 6,285 people visiting sites (2009–2011)	Need to demonstrate impact of innovation to Central Government funders

(*Source* Author)

resources. Reflection of socio-spatial development practice to evaluate and disseminate learning typically occurred at the end of each case study, or significantly, at the end of each cycle of development if the case underwent multiple cycles (e.g. Southsea Greenhouse, and The Charles Dickens Orchard Trail, Portsmouth). Finally, this learning and reflection can cause reconfiguration of delivery of spaces and services in order to advance or sustain the initiative.

In providing opportunities for social empowerment, the cases demonstrate how this is dependent on prior identification and development processes. Reflection activity typically happens after provision activity. The basic sequence of social innovation agency is shown in Fig. 12.1. This basic pattern features 'spatiotemporal' dependent events and outcomes, relating to both space and time, as important components associated with key innovation processes to support this sequence. The event is the sum of contextual inputs (e.g. sociocultural, political, economic, and geographical) and is the trigger for social innovation action. Outcomes are observable social

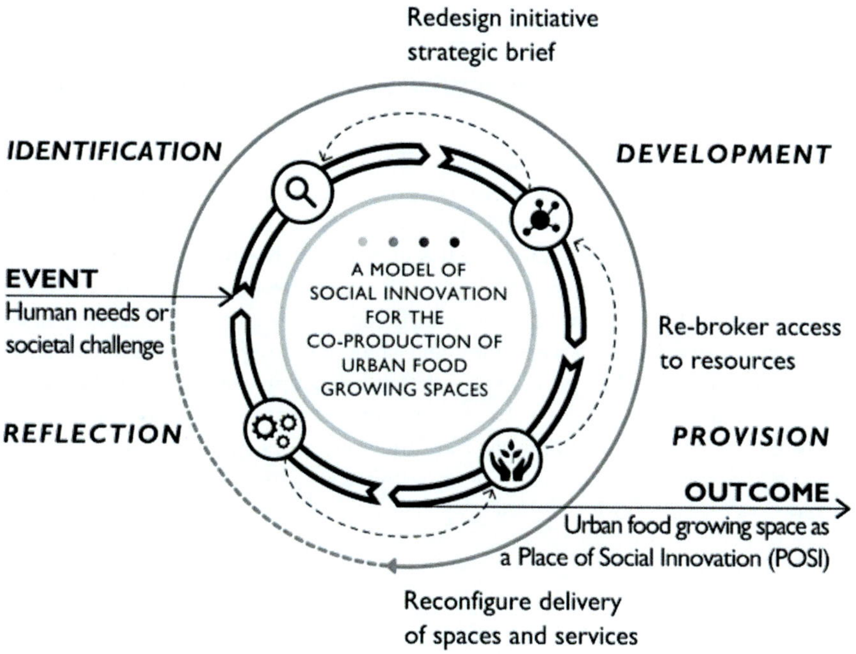

Fig. 12.1 Basic pattern identified in key social innovation processes (*Source* Author)
Note Dashed arrows between key processes indicate the revisiting of prior processes in order to progress innovation

and spatial changes that arise from examining cases over time; they may be short-term or long-term, planned or accidental, and different in influence. In this regard, the case study outcomes were shown to be multiple and varied in their production of urban food spaces.

The basic pattern corresponds to a general explanation of events, activities, and interactions for each case; this incorporates the main evolving pattern of social innovation action without alteration to its fundamental nature. Through evaluation of process research measures for the development of effective theory as proposed by Langley et al. (2013), Pentland (1999), Van de Ven and Poole (2005) (see Chap. 2), the pattern exhibits generality in the extent that it corresponds to a varied group of cases, simplicity in possessing few components or attributes, though potentially limited accuracy as the pattern only mirrors case narratives through broader components. Therefore, each UK-based case study is visualised in the following section to interrogate and advance this pattern. Aside from cross-examining the pattern it will offer a basis to explain the reason the basic pattern occurs.

12.2 Visual Analysis of Identified Social Innovation Pattern

I mapped case narratives across six case studies (see Chap. 2). Chronological events, activities, and decisions pertinent to each case were positioned within analytical scales. Key social innovation processes identified were positioned over the case events (Chaps. 6–11). For each case, the map establishes the pattern of initiative development. Figures 12.2 and 12.3 visually map these processes across two UK cities, Brighton & Hove and Portsmouth. The figures are smaller representations of maps featured within each case study and combined to enable assessment across cases. My intent is not to fully display the events within each case (as this can be examined in detail within Chaps. 6–11), but rather to observe basic patterns to social innovation in the production of urban food growing spaces. Figure 12.4 combines the visual narrative maps of the case studies within each city to further highlight these patterns.

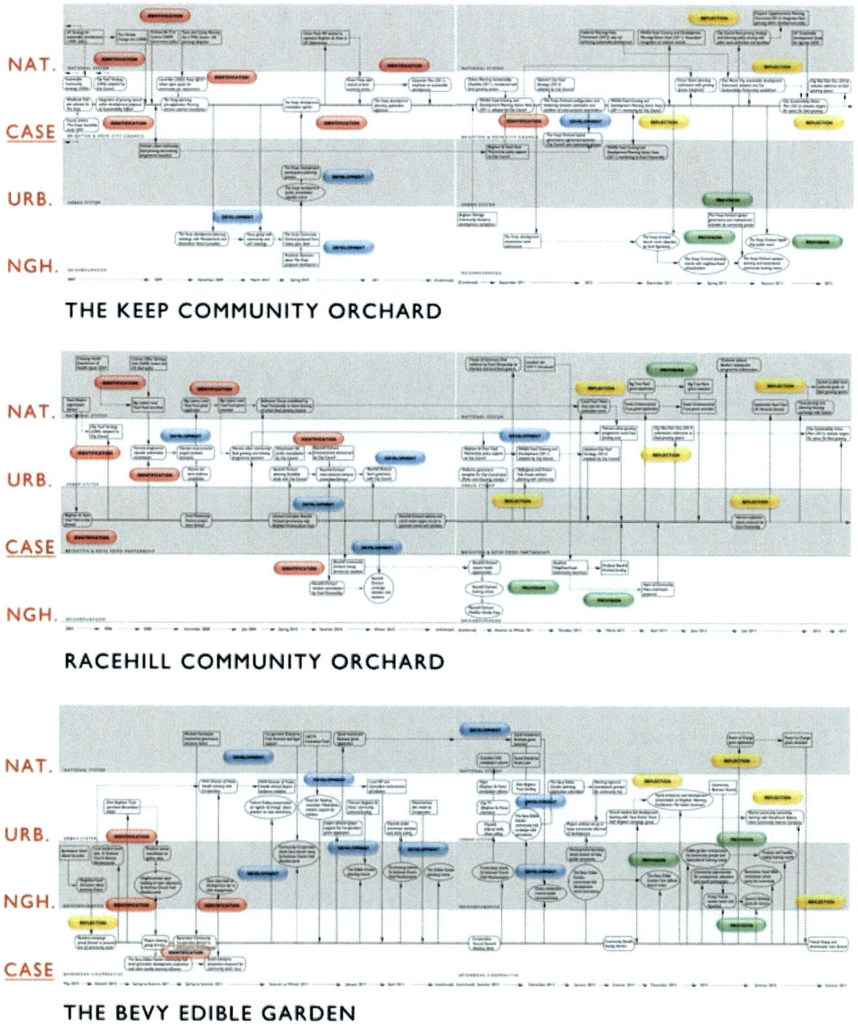

Fig. 12.2 Key social innovation patterns for Brighton & Hove cases of urban food growing (*Source* Author)
Colour-coded polygons represent groupings of key processes: Identification of human needs or societal challenges to sustainable development (Red); Development of social relations in systems or structures (Blue); Provision of opportunity for social empowerment (Green); Reflection of socio-spatial development practice (Yellow).
Contextual scales are abbreviated as: National System (NAT.); Urban System (URB.); Neighbourhood (NGH.); Focal case stakeholder (CASE).

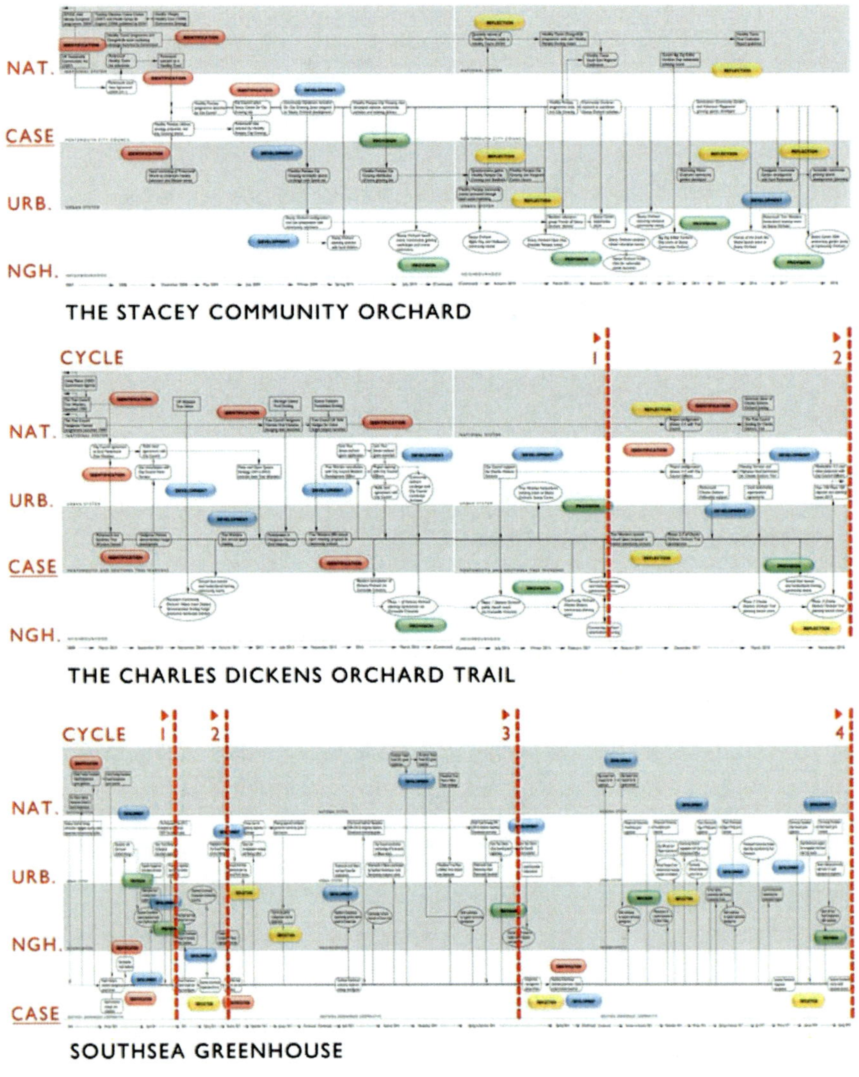

Fig. 12.3 Key social innovation patterns for Portsmouth cases of urban food growing (*Source* Author)

Colour-coded polygons represent groupings of key processes: Identification of human needs or societal challenges to sustainable development (Red); Development of social relations in systems or structures (Blue); Provision of opportunity for social empowerment (Green); Reflection of socio-spatial development practice (Yellow).

Contextual scales are abbreviated as: National System (NAT.); Urban System (URB.); Neighbourhood (NGH.); Focal case stakeholder (CASE).

Dashed vertical lines indicate numbers of cycles of social innovation within cases.

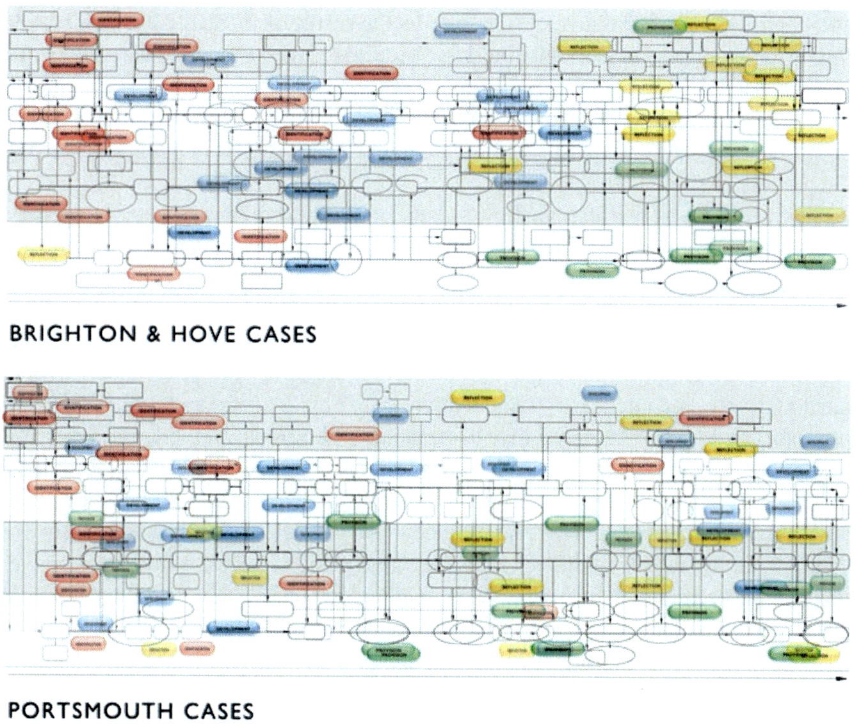

Fig. 12.4 Key social innovation patterns for Brighton & Hove and Portsmouth, UK cases of urban food growing (*Source* Author)
Three cases are combined and overlaid for each city to highlight innovation patterns.
The Brighton & Hove cases in general correspond to the basic pattern in the sequence of social innovation, albeit with a small degree of overlap between processes.
The Portsmouth cases show some variation in the ordering of their process sequence due to the presence of multiple cycles of social innovation, causing earlier processes to be revisited in order to progress innovation.

12.3 Conclusions: Patterns to Key Social Innovation Processes

Several conclusions are drawn from my analysis of social innovation processes and patterns.

First, the basic pattern of social innovation identified captures the detail of case action in addition to a general account of activity and relationships.

Second, this sequence does not automatically move or operate in a single direction, rather multi-directional movement is detected. Changing environmental conditions suggest that key innovation processes are potentially adaptable depending upon circumstance. My finding supports Sztompka (1993) and Pettigrew (1997) in that social processes can be linear, directional and cumulative, while others can be non-linear, radical and transformational. In all cases, ineffective realisation of local needs and development of social relations typically causes future challenges to the growth and continuation of initiatives. This may be the result of inadequate consideration, timing or manoeuvring of stakeholders. Moreover, this situation necessitates the revisiting key processes (e.g. Southsea Greenhouse, Portsmouth) or it may thwart their long-term operation altogether (e.g. The Keep Orchard, Brighton & Hove). I propose an openness to multi-directional movement as therefore being important to the process of social innovation.

Third, the maps and explanation of this basic pattern emphasise various flows of social action inside innovation cycles and stages of development. This indicates that key social innovation processes may be performed consecutively. For example, Southsea Greenhouse, Portsmouth attempted to identify a new site and develop relationships to access material resources while providing community activities. Equally, The Bevy, Brighton & Hove progressed in conjunction with stakeholder involvement in developing another community business. This is accounted for in innovation processes being undertaken relative to the actions of various individuals and groups in a dynamic local environment. However, various courses of action also indicate this basic sequence is seen working in larger scale sequences, e.g. both the Racehill Orchard, Brighton & Hove and Stacey Orchard, Portsmouth were planned as demonstration projects within specific areas though connected to wider urban development programmes. My finding also implies the basic sequence may again be observed at subtler analytical scales, such as the household, or in shorter weekly event chronologies, and indicates the suitability of the innovation pattern to a more extensive application.

Fourth, key social innovation processes are exemplified by their constant state of change. Additional actions in the identification of needs and development of social relations might be performed throughout spatial development. The basic pattern is an 'ideal type' of key social innovation processes in the production of urban food growing spaces, and thus frames the pattern found in a model situation. As such, the pattern underscores the influence of earlier processes on those subsequently performed and therefore comprises the primary reason for the presence of the patterns. Future social innovation processes advance and at the same time require components of previous processes.

Fifth, there are positive 'feedback loops' evident in case activity in which a change causes additional changes, particularly in longer-term learning processes, case developments and outcomes (see Fig. 12.5). This repeated method is characterised by Carless (2019) as double-loop learning or 'feedback spirals', being temporal and repetitive, in the cyclic re-evaluation of problem-solving approaches. In this regard, urban stakeholders had to gather new evidence, subsequently adjust values, and modify development practice based on new knowledge. This process may occur internally within the focal stakeholder group or externally through the dissemination of project learning to other localities. The presence or absence of active learning is a contributing factor to case development. My finding supports insights from social innovation research (e.g. SIX 2010) regarding feedback loops between key innovation processes and fostering an ability to learn being significant to the realisation of initiatives. In particular, as initiatives become more complex the integration of

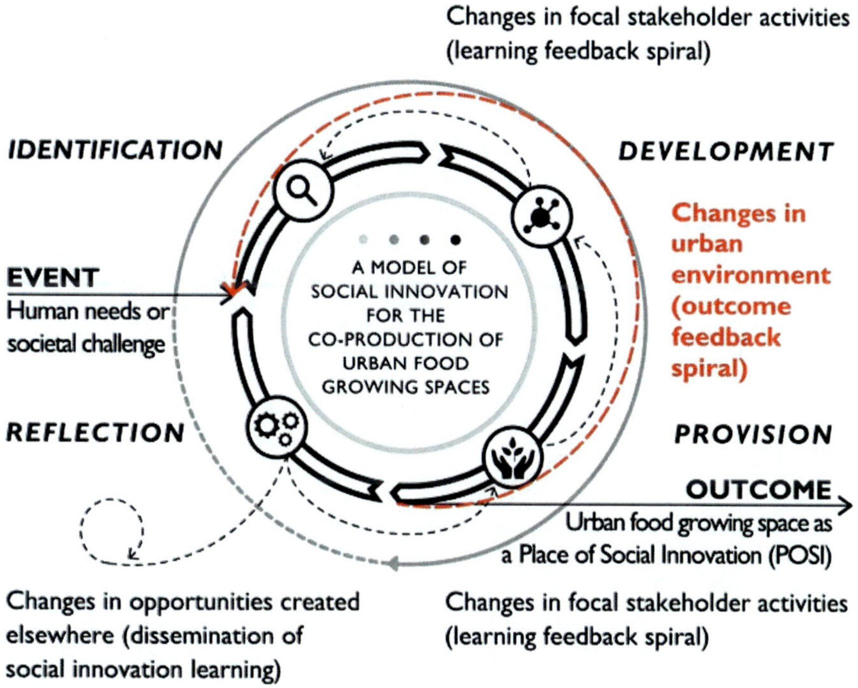

Fig. 12.5 Feedback spirals affecting case activity and outcomes changing the urban system to cause future environmental scenarios for social innovation (*Source* Author)

reflection and learning activities by project stakeholders is necessary to advance innovation.

Sixth, cycles of innovation agency are also observable within the basic pattern, connecting outcomes to reflection activity and the knowledge gained from this reflection process leads to further identification of needs by innovation stakeholders. Outcomes also relate to social and material changes in the urban system and therefore changes in environmental scenarios for further social innovation activity. Here, Pettigrew (1997) reminds us that 'context is not just a stimulus environment but a nested arrangement of structures and processes where the subjective interpretations of actors perceiving, learning, and remembering help shape process' (p. 341). As such, processes across the cases are both progressed and restricted by features of context whilst also shaping local contexts. My finding substantiates understanding of social action in the 'production of space' (e.g. Lefebvre 1991; Soja 2010), regarding innovation both shaping and being shaped by the urban environment and the intensity of its social interactions.

In summary, the basic pattern of social innovation established in my analysis both verifies and advances my study's conceptual framework (Chap. 4). Essentially, the provision of opportunities for social empowerment requires components of identification of needs and development of social relations; indicating that in order to develop these distributive spaces will require earlier identification and development action. On its own, this basic pattern indicates an 'ideal type' of sequence of social innovation around timely and appropriate action for the production of urban food growing spaces. While beneficial to understand the internal forces of key innovation processes it is recognised that this presents only part of the account of spatial production by urban stakeholders. A further understanding of stakeholders' agency on innovation development requires the positioning of case action within a structured environment of multiple spatial scales. In addition to the agency of case stakeholders documented through mapping and assigned to key innovation processes, there is a need to consider the various contextual influences on this social action across space and time (see Chap. 13).

References

Carless D (2019) Feedback loops and the longer-term: towards feedback spirals. Assess Eval High Educ 44(5):705–714

Langley A, Smallman C, Tsoukas H, Van De Ven AH (2013) Process studies of change in organization and management: unveiling temporality, activity, and flow. Acad Manag J 56(1):1–13

Lefebvre H (1991) The production of space. Blackwell Publishing, Oxford

Pentland BT (1999) Building process theory with narrative: from description to explanation. Acad Manag Rev 24(4):711–724

Pettigrew AM (1997) What is a processual analysis? Scand J Manag 13(4):337–348

Poole MS, Van de Ven A, Dooley K, Holmes M (2000) Organizational change and innovation processes: theory and methods for research. Oxford University Press, Oxford

SIX (2010) Study on social innovation. The Young Foundation, London

Soja E (2010) Seeking spatial justice. University of Minnesota Press, Minneapolis

Sztompka P (1993) The sociology of social change. Blackwell, Oxford

Van de Ven A, Poole M (2005) Alternative approaches for studying organizational change. Organ Stud 26(9):1377–1404

Chapter 13
Social Innovation Influences and Contextual Dynamics in the Production of Urban Food Spaces

Abstract This chapter draws together observations from across the cases studies to address the research question: 'How do contextual forces across spatial scales influence the social innovation actions of urban stakeholders in the production of food growing spaces?' The chapter identifies contextual influences on this activity across three spatial scales: neighbourhood, urban system, and national system. For each scale, influences are categorised into groupings of principal forces to support explanation of their effect on social innovation. The findings show that in general urban stakeholders have variable authority and means to innovate, and their actions are progressed and restricted by various contextual forces within a social innovation environment of multiple scales.

Keywords Social innovation processes · Contextual forces · Spatial scales · Socio-spatial dynamics · Urban food spaces · Co-production of space · Social innovation · Process research

13.1 Contextual Forces Across Spatial Scales Influencing Social Innovation

It has been widely suggested that social innovation agency is 'contextually contingent' (see, e.g. Baker and Mehmood 2015; Howaldt et al. 2017; Pisano et al. 2015). Nevertheless, understanding of the context and dynamics affecting how social innovation evolves is underdeveloped. Current studies on social innovation in urban spaces have paid limited attention to influencing factors across various spatial scales; presenting outstanding questions around stakeholder action and who exactly is producing urban space. I therefore decided in my research to focus on three particular levels of urban stakeholders, which has supported an understanding of the interactions between social agency and innovation environments.

In my research, the narrative account of each initiative's development illustrates the context specific progression of social innovation. As such, various contextual forces are seen to affect key processes in how urban stakeholders identify needs, develop social relations, provide opportunities, and reflect on development practice in the production of food growing spaces. I will now advance the recognition of

N. Ardill, *Growing Food in Cities*, Cities and Nature,
https://doi.org/10.1007/978-3-030-98475-5_13

stakeholder actions as being dependent on context by identifying the principal forces across spatial scales that affect social innovation. The following analysis builds on the identification of patterns to key social innovation processes (see Chap. 12).

Through an in-depth analysis of six case studies across the two cites of Brighton & Hove and Portsmouth, UK over a sustained period of time (Chaps. 6–11), 124 influences were identified on how urban stakeholders' actions were affected by contextual forces across three spatial scales: neighbourhood, urban system, and national system. Nearly half of these forces were derived from the urban system (45%) with the remainder split across the neighbourhood (30%) and the national system (25%). The division of three contextual scales is for analytical reasoning with scales interconnected in application. For each scale, influences are categorised into groupings of principal forces to support explanation of their effect on social innovation.

13.2 Neighbourhood Scale

There are four identifiable categories of neighbourhood forces shaping social innovation in the case studies: (1) Interest and recognition; (2) Socio-political history; (3) Experience and skill; (4) Existing infrastructure, resources, and assets. Table 13.1 summarises these forces and their respective influences.

First, the degree of 'Interest and recognition' among local residents and has a significant influence on how innovation developed. This mirrors an important characteristic of a neighbourhood in the degree of interest residents have in area matters, including the ability of focal stakeholders to communicate these issues and propose spatial interventions in a way that is accepted by the target neighbourhood. The case studies show that for initiatives to be effective (e.g. community buy-in and long-term continuation) they must realise shared values among neighbourhood residents to reach their target group, whilst restrictions and obstacles such as disinterest or lack of participation must be overcome. My observation is supportive of the claims of Manzini (2015) and Moulaert et al. (2010) around social innovation in that neighbourhoods are 'active agents' of social change (see Chap. 3).

Second, the 'Socio-political history' and past events of the neighbourhood influence the ability of focal stakeholder to activate local people, and to communicate, collaboratively design initiatives, and develop spaces (e.g. The Bevy, Brighton & Hove and Southsea Greenhouse, Portsmouth). This factor relates to the provision of opportunities for social empowerment, and the significance of shared values to forming a culture of self-help in building 'collective capabilities' to create the environmental conditions for social justice in the development of more equitable cities (as suggested by Fainstein 2010 and Marcuse 2009).

Third, and building on my previous observation, neighbourhood attributes of 'Experience and skill' in several areas; spatial development, communications, and entrepreneurship affected the realisation of stakeholder attempts to configure provision strategies for their projects. This characteristic influenced the feasibility of focal stakeholders to engage and utilise neighbourhood participants. Furthermore, it not

Table 13.1 Neighbourhood forces with influence on social innovation action

Force	Influence on cases in Brighton & Hove and Portsmouth, UK (Case study references shown abridged in brackets)
Interest and recognition	• Willing resident volunteers in initiatives and attraction to community food growing practices (Variable in all cases) • Enterprise community share issue and co-operative membership, supports community buy-in (The Bevy); (Southsea) • Friends' groups devolve spatial governance and support long-term continuation of initiative (The Bevy); (Racehill); (Stacey) • Pursuit of collective spatial governance to support neighbourhood development (The Bevy); (Southsea) • Community food growing space ceases public access (The Keep) • Group split, reduced access to space, services and devolved spatial governance (Southsea)
Socio-political history (past events)	• Mistrust and/or lack of confidence in local authority to provide services and facilities required to meet local needs (The Bevy); (Southsea) • Urban decline in the neighbourhood (The Bevy); (Southsea)
Experience and skill	• Prior green space development within the target area. Knowing what works or does not work in a community and authority and trust gained amongst residents (The Bevy); (Racehill); (Dickens); (Stacey) • Experience in urban regeneration, community development and communications (to gain support). Supported understanding of urban system dynamics and approaches to development funding (Racehill); (Southsea); (Stacey)
Existing infrastructure, resources, and assets (physical and social)	• Availability of unexploited public land (Racehill); (The Keep); (Southsea); (Dickens); (Stacey) • Former public house in neighbourhood available for community tender (The Bevy) • Neighbourhood community centre that is regularly used by residents and target demographics (Stacey) • Temporary relocation to nearby privately owned business premises, material and moral support facilitated enterprise development when facing closure (Southsea) • Initiative configured around identified needs and opportunities, participatory planning, focus groups and meetings (The Bevy); (Racehill); (The Keep); (Dickens) • Limited consultation of others in initiative planning (Southsea); (Stacey)

Source Author

only affected bottom-up grassroots initiatives, but also the capacity of intermediate level non-governmental organisations and top-down local authority approaches to urban development (e.g. Racehill Orchard, Brighton & Hove and Stacey Orchard, Portsmouth). In particular, community development experience and knowledge of practicable methods within a target neighbourhood was found to be significant.

Fourth, the capacity to use and multiply 'Existing neighbourhood infrastructure, resources, and assets' (both physical and social) for initiatives was shown to influence their development by creating resource opportunities and resolving challenges. The concentrated focus was on the assets and strengths of a neighbourhood rather than its deficits, to help satisfy residents' needs. The case studies demonstrate that material resources in public spaces or enclosed spaces can support 'asset-based planning' in neighbourhood interventions. Additionally, consultations in social interactions occurring between initiatives and residents influenced the effectiveness of these asset-based approaches by using local insight to enhance neighbourhood satisfaction and spatial outcomes.

These four neighbourhood forces affect each social innovation process arising and are especially seen across provision activity when obstacles happen. For example, Southsea Greenhouse, Portsmouth was unable to sustain a trading space due to inadequately identifying needs (and thus potential solutions) and not establishing enough neighbourhood interest to make it financially viable. The stakeholders required assistance from a local business to permit necessary time to reconfigure their enterprise. Subsequently, the enterprise evolved into a community food growing initiative; presenting a more sustainable model that was widely accepted and used by residents. The case can be understood as the neighbourhood not accepting the innovation at first and asserting its presence by not offering compatibility until reconfiguration. My observation here supports Moulaert et al. (2010) in underlining the significance of the community relationship and contact of focal stakeholders to a neighbourhood to counter social innovation constraints and obstacles (see Chap. 3). Consequently, an enhanced understanding of the target beneficiaries of initiatives means increasing the opportunity for both identifying neighbourhood needs and the capacity for responding more effectively to these needs.

My last finding signifies a broader theme across the case studies. Characteristically, focal stakeholders sought to introduce a solution for meeting local needs that was cooperatively advanced with their target neighbourhood (albeit with varying degrees of resident participation). The 'Healthy Pompey' programme (Stacey Orchard, Portsmouth), however, does not fit within this observation; being planned independently of its target neighbourhoods in a top-down manner. Portsmouth City Council attempted to transfer their predetermined strategy for developing food growing practices locally, whilst selecting several neighbourhoods they deemed appropriate. This is the most irregular example across the six case studies as the local authority exhibits such an unattachment to the significance of neighbourhood forces. However, it is acknowledged that all the cases studied, and their associated stakeholders did this to a certain degree, owing to the need to make choices within time and human resource constraints but also when they are mainly inspired by a single individual's vision (e.g. Southsea Greenhouse, Portsmouth).

13.3 Urban System Scale

Four categories of urban system forces are identified as affecting the development of social innovation: (1) Urban politics and policy; (2) Professional services and collaborations; (3) Intermediation; (4) Innovative environment. Table 13.2 summarises these forces and their respective influences.

First, 'Urban politics and policy', such as city leaders and development plans can cause and accelerate initiatives; offering support to urban stakeholders in stimulus and funding. Additionally, this force presents several potential resources: (1) Cultural in encouragement, indirect backing, and approval; (2) Societal in authority and promotion via local councillor association; (3) Material in access to public land; and (4) Human and organisational in professional expertise, coordination, and steering. I will explain further here the significance of urban policy on material resources. Public land is overseen in the built environment by statutory authorities and consequently their land-use policy, concerning open space, urban food growing, and sustainable development, influences the way urban stakeholders may or may not access public resources (e.g. Racehill Community Orchard, Brighton & Hove). In combination the forces are evident in affecting any recognised openings for urban stakeholder agency, though they also affect each key social innovation processes. They affect how focal stakeholders identify needs and interpret opportunities, the means and setting in which stakeholders develop relations in negotiating resources and backing, how provision spaces are realised, and the nature of reflection activities internally and with others.

Urban policy can generate social obligations and conditions. An observable requirement is in the planning process of The Keep Orchard, Brighton & Hove. The local planning authority performed a significant function in interpreting national planning policy frameworks, specifically in the provision of public open space as a planning obligation to meet the requirements of their Local Development Framework. Equally, both The Bevy, Brighton & Hove and Southsea Greenhouse, Portsmouth had to negotiate urban planning as their initiatives required consents. As local authorities translate national planning frameworks into the urban context, these frameworks are adaptable, developing across time and being influenced by various local participants and actions.

What is not so well known however, is how urban policy may generate impetus behind spatial practices and their adoption within the urban environment. This dynamic can support the innovation actions of urban stakeholders. An example is the shifting position of Brighton & Hove City Council towards urban food growing, moving from a passive stance in 2003 to developing in 2006 a Sustainable Community Strategy and Food Strategy; establishing a strategic vision to increase land available for food growing. This opened-up opportunities for local organisations, including Brighton & Hove Food Partnership and Brighton Permaculture Trust to develop community orchards in public areas. Opportunities were further enhanced in 2011 through a change of city administration, advancing policy that integrated food growing within planning and sustainable development. In this regard, political

Table 13.2 Urban system forces with influence on social innovation action

Force	Influence on cases in Brighton & Hove and Portsmouth, UK (Case study references shown abridged in brackets)
Urban politics and policy	• Local Plan (2005) Policy QD20 for urban open space. Led to City Council recommendation to include open space within development masterplan for community use (The Keep) • Portsmouth Local Area Agreement (2008–11). High obesity rates, especially people living inactive lives in deprived wards, demonstrated need for innovative health initiatives (Stacey) • Brighton & Hove Food Strategy (2006; 2012). Set urban food growing targets (Racehill); (The Keep) • Sustainable Community Strategy (2006; 2010). Set urban food growing targets (Racehill) • City Council provide resource opportunities. Project stimulus and funding (The Keep); (Stacey) • Access to public material resources. Public land and 'peppercorn lease' agreements (Racehill); (Southsea); (Dickens) • Green Party elected to local authority power. Stimulated local sustainable development policy strategies and adoption (The Keep) • Legal support in conditional planning approval granted for enterprise: Permitted community hub (The Bevy); Restricted enterprise to produce trading only (Southsea) • Supportive City Council Officers as institutional 'gatekeepers'. Assistance and recognition given (Southsea); (Racehill); (Dickens) • Unsupportive City Council Officers. Lack of recognition or restriction to innovation (Southsea)
Professional services and collaborations	• Co-design with experts to configure initiative (The Bevy); (Dickens) • Formation of cross-sector collaboration and service delivery partnerships (The Bevy); (Racehill); (Southsea); (Stacey); (The Keep) • Cross-sector partnership falter due to lack of accountability of City Council in land management responsibility (The Keep)
Intermediation (people and spaces)	• Intermediary agents and spaces facilitate, configure and broker relationships to further support and development (The Bevy); (Racehill); (The Keep); (Dickens)

(continued)

Table 13.2 (continued)

Force	Influence on cases in Brighton & Hove and Portsmouth, UK (Case study references shown abridged in brackets)
Innovative environment	• Supportive network of individuals, community groups, and local businesses (The Bevy); (Racehill); (Southsea); (Dickens); (Stacey) • Creates impetus to initiatives novel to their context (Southsea); (Dickens) • Public and media events. Spreads awareness to support engagement with initiatives and participant recruitment (The Bevy); (Racehill); (Southsea); (Stacey) • Small grants programmes. Facilitate service delivery, set guidelines for community development and neighbourhood transformation (The Bevy); (Racehill); (Southsea); (Dickens) • Shared development learning with resident groups and enterprises within urban system (The Bevy)

Source Author

stakeholder positions typically shift across the evolution of innovation, e.g. through local elections and individual changes in personnel, presenting openings though also potential challenges in consistency. The interests of political entities are therefore not fixed but apply changing effects over time.

Second, 'Professional services and collaborations' at the urban system scale, influenced how initiatives were planned, designed, and delivered. This collaborative city shaping, and alternative urbanism works across urban divides, between citizens and the state, the informal and formal. Lay people and professionals here were actively cooperating in developing spatial experiments and innovations (e.g. Dickens Orchard, Portsmouth). This force overcame urban obstacles in deficits in local knowledge and skills in both directions. The cases show that to resolve multiple interests, which can act in pursuit of conflicting urban plans, cross-sector collaborations, partnerships, and development coalitions are effective. However, a lack of individual accountability or uneven arrangements in state-civil partnerships may present barriers and disempowerment (e.g. The Keep Orchard, Brighton & Hove).

Third, the influence of innovation 'Intermediation' across the urban system, linking people and resources, supported the development of cooperative relationships in the forging of alternative urban development scenarios. For instance, in The Bevy and Racehill Orchard, Brighton & Hove, this social agency increased citizen access to local decision-making and worked to include marginalised community voices in urban governance.

Fourth, the respective urban systems revealed qualities of an 'Innovative environment', seen by Klein (2009) and MacCallum et al. (2009) as important to urban development and social innovation. This force is understood as an environment in which groups of sociocultural relationships in geographical proximity support the advancement of social action. My observation is perhaps predictable, considering Brighton & Hove's reputation as a hub of activism (see Chap. 2). However, less recognised is the influence of Portsmouth's social environment in contributing to the emergence of innovations. Locally, the forces exerted can be observed though equally understated in giving backing to and creating impetus behind certain initiatives novel to their context (e.g. Southsea Greenhouse and Dickens Orchard, Portsmouth). This provided encouragement, indirect support and approval whilst stimulating extra confidence and motivation in stakeholder actions. Moreover, these innovative environments created opportunities and spaces for the emergence and growth of social innovation, in which proposals could be articulated, shared and developed (e.g. The Bevy, Brighton & Hove), and substantiates the argument of Adams and Arnkil (2013) in cities being catalysts and brokers of social innovation.

Overall, the influence of the urban system forces is significant as they contextualise innovative actions and position multilevel stakeholders, in a spatial and political arrangement, within networks and interest groups. The urban system also moderates opportunities and barriers caused from the national system to contextualise innovation forces, e.g. socio-ecological action in land-use planning is dependent on local attitudes and approaches. Likewise, the agency of national funding bodies and advocacy organisations transmit into urban forces to affect openings and the configuration of localised initiatives. In this sense, there is a rational consistency with the national system forces; the urban system is connected to and situated within the larger innovation system, and consequently underlying national forces become apparent at the urban scale.

13.4 National System Scale

Three categories of national system forces shaping social innovation can be identified across the cases: (1) Government policy and strategy; (2) Funding bodies and grant making trusts; (3) Advocacy organisations and networks. Table 13.3 summarises these forces and their respective influences.

First, 'Government policy and strategy' was shown to have an impact on the way urban stakeholders identified needs, developed social relations, provided opportunities, and reflected on development practice, as demonstrated by recurring effects amongst the cases. National strategies set objectives for societal and urban change, whilst subsidising localised agency. Conversely, austerity politics and local authority cuts influences grassroots action to cover service deficits, e.g. The Bevy, Brighton & Hove and Southsea Greenhouse, Portsmouth. The development of strategic urban policies such as Healthy Towns was another innovation prompt. In three case studies government policy was directly responsible for stimulating the production of public green spaces (e.g. Racehill Orchard and The Keep Orchard, Brighton & Hove and Stacey Orchard, Portsmouth). However, government policy is also regulated, making demands on stakeholders to modify development templates (Racehill

Table 13.3 National system forces with influence on social innovation action

Force	Influence on cases in Brighton & Hove and Portsmouth, UK (Case study references shown abridged in brackets)
Government policy and strategy	• Healthy weight, healthy lives: a cross-government strategy for England (2008). 'Healthy Towns' programme (2008–2011). Change4Life. Set target to tackle obesity levels in urban environments via residents' lifestyle and behavioural changes (Stacey) • Town and Country Planning Act (1990) Section 106 agreement planning obligation. Facilitated spatial 'planning gain' in approval (The Keep) • The Climate Change Act (2008). Targeted the reduction of UK carbon emissions by 80% by 2050 with country's food system a significant area to be addressed (Racehill); (The Keep) • UK Sustainable Communities Act (2007). Set target for local authorities to reverse urban trend of community decline in ongoing loss of local facilities and services (Stacey) • Localism Act 2011, devolution of decision-making powers from central government control to individuals and communities. Influenced experimentation with participatory processes (Racehill) • Co-operative and Community Benefit Societies Act (2014). Influenced governance structures (The Bevy); (Southsea) • Living Places: Cleaner, Safer, Greener (2002) programme. Backing for national Tree Warden Scheme (Dickens)
Funding bodies and grant-making trusts	• Big Lottery 'Local Food' Fund. Created opportunity, set guidelines for shared learning to other localities (Racehill) • Social Investment Business Foundation. Created opportunity, set guidelines for community business (The Bevy) • UnLtd Funding Foundation for Social Entrepreneurs. Created opportunity, set target for enterprise to support urban development (Southsea) • Veolia Environmental Trust (Landfill Communities Fund). Facilitated service delivery (Racehill) • The People's Health Trust. Facilitated service delivery, set guidelines to address health inequalities in deprived areas (Racehill) • Big Tree Plant Government campaign. Facilitated service delivery, set guidelines to support disadvantaged communities to establish and protect green spaces (Racehill) • EC European Social Fund. Facilitated service delivery, set target for community development to support employment opportunities through innovation (Southsea) • Big Lottery Fund Awards for All. Facilitated service delivery, requirement for inclusive programme to support social cohesion (Southsea)

(continued)

Table 13.3 (continued)

Force	Influence on cases in Brighton & Hove and Portsmouth, UK (Case study references shown abridged in brackets)
Advocacy organisations and networks	• Food Matters. Political lobbying support and coordination between national and local strategies (Racehill) • Sustainable Food Cities Network. Supported nationwide dissemination of innovation learning (Racehill); (The Keep) • Tree Council UK. Coordination between national and local agendas', grant funding, planting expertise and fruit tree provision (Dickens) • Communication channels aid dissemination of learning to other localities (The Bevy); (Racehill); (The Keep); (Southsea); (Stacey) • Learning from resident groups in other localities influences planning and development of neighbourhood initiative (The Bevy)

Source Author

Orchard, Brighton & Hove), or adjust to fluctuating political environments (Stacey Orchard, Portsmouth).

Observing social innovation patterns as being shaped by public policy is not especially revealing or unexpected. Yet, it confirms in practice Mulgan's (2006) claim on the process of social innovation concerning systematic approaches or 'innovation accelerators' affecting the evolution and spreading of collaborative development. It informs the way policy influences though also the reason it performs a necessary function and thus policies are recognised as causing various effects on innovation.

Second, 'Funding bodies and grant-making trusts' comprised a significant category of influences on social innovation from the national system. This force is interlinked with government policy. Bodies such as the UK Big Lottery Fund gave projects funding in line with objectives set by government. A prevalence of social enterprise and entrepreneurship funds and social financing organisations were also found. Alongside government policy, these forces occur at the urban to neighbourhood scales. For example, the rise of localism politics and the 'Big Society' ideology following the 2007–08 financial crisis led to the UK Localism Act 2011. This approach positioned citizens and community business as taking over and running 'neighbourhood assets' (buildings and open spaces) and providing local services and facilities on behalf of local authorities. Contextually, it meant that several national organisations were given government funding at the time; creating the opportunity to develop bottom-up enterprises (e.g. The Bevy, Brighton & Hove and Southsea Greenhouse, Portsmouth), demonstrating how a particular national policy affects innovation.

By examining the influence of government policies on public expenditure and innovation funding, they can be differentiated as being 'distributive' and 'redistributive'. These goal-orientated approaches frame and support innovation differently, thus influencing different local dynamics, especially in local authority action. For example, the Climate Change Act 2008, as distributive policy, sought reduction of carbon emissions. This influenced localised sustainable development and formulation of planning policy, using public funds and resources to stimulate urban green infrastructure to benefit the wider society (e.g. The Keep Orchard, Brighton & Hove). Whereas, Healthy Weight, Healthy Lives: a Cross-Government Strategy for England (2008) typifies a redistributive approach in targeting specific demographics in families and children at risk of obesity, especially those from within disadvantaged communities, to improve individual healthy choices and behaviours. Consequently, a particular section of society and targeted geographical area benefitted from publicly funded neighbourhood interventions (e.g. Healthy Pompey and Stacey Orchard, Portsmouth).

Third, and the final category of national influence, is 'Advocacy organisations and networks'. Organisations affiliated to urban stakeholders were significant bridges between national agenda setting or problem identification and localised implementation of solutions. They contributed to the understanding of societal challenges by urban stakeholders within subsidiary organisations, helping them to interpret and contextualise local needs within the broader national environment, e.g. Racehill Orchard and The Keep Orchard, Brighton & Hove and Dickens Orchard, Portsmouth. This force also influenced stakeholders' reflection activities in how they evaluated initiatives, and subsequently, the sharing of learning generated with other localities, such as where strategies were successful or not in meeting their aims. Here, the availability and networked outreach of multiple communication channels such as the UK Sustainable Food Cities network affected opportunities for distribution of urban development knowledge, and thus potential for diffusion and replication of innovation elsewhere to increase coverage of innovation solutions.

Altogether, these three qualities of the national system are detected as underlying influences on stakeholders' innovation agency. They manifest across all six UK-based cases, yet with varying impact. Observation of the effects of government policy and strategy typically increases in the top-down case studies compared to others. This is ascribed to the true state of affairs (or 'how things are in the world'), though also to my research design; the selection of local authority cases meant that the influence of central government was more likely to be a significant factor on this stakeholder. Conversely, these variations reflect the chosen strategies of each local authority to addressing identified needs and their specific contexts. Additionally, it is possible to identify the influence of public policies on various funding bodies and grant-making trust agendas, and the type of innovation action enabled at the grassroots.

13.5 Conclusions: The Influence of Contextual Forces on Social Innovation

Based on my analysis of contextual forces, some additional conclusions on social innovation action can be made.

First, urban stakeholders have variable authority and means to innovate, and their actions are progressed and restricted by contextual forces at various spatial scales. The national system principally functions as underlying effects, largely outside the bounds of manipulation of stakeholders though causing openings for localised innovation. The urban system contextualises national system forces according to the characteristics of the urban environment (e.g. political, material, and sociocultural) and positions innovation stakeholders locally. These urban stakeholders have substantial, though varying, power on coordinating changes at this scale. This is attributed to the innovation process being influenced by various urban participants across civil society, public and private sectors. What is feasible or unfeasible for innovation stakeholders is conditional on the encouragement of these individuals, groups, and local institutions. Lastly, the neighbourhood influences potential sequences of social innovation in an understated though significant way.

Second, the examination of targeted neighbourhood interventions implies querying the extent urban stakeholders identified local needs alongside their intended beneficiaries. While stakeholders' initiatives were typically aimed at a particular neighbourhood, it was shown that their understanding of these local areas in terms of residents' wishes, or requirements was not always the case. In some instances, stakeholder relation and contact to the intended neighbourhood was indirect or largely unattached (e.g. Southsea Greenhouse, Portsmouth or the wider 'Healthy Pompey' programme of Stacey Orchard, Portsmouth). Stakeholders here had to gain experiential knowledge of the neighbourhood; consequently, facing additional challenges to developing urban food growing initiatives. The detachment of stakeholders instigating spatial interventions from the intended neighbourhood beneficiaries signifies these local residents as just one of numerous, wider groupings of urban participants that innovative agency aims to galvanise, with community organisations, institutions and statutory authorities proposed as important groups. Therefore, stakeholder understanding and contact to a target neighbourhood should not be taken for granted though equally it is significant to the evolution of practicable and sustainable initiatives.

Lastly, in order to be effective, urban stakeholders' innovation activity must be adjustable to potential restraints to development or utilise shifting contextual forces and openings. The process of social innovation is 'spatiotemporal' dependent, in relating to both space and time. The cases illustrate how contextual forces develop and various participants' roles will often fluctuate. Adaptability is thus a useful quality as focal stakeholders acquire knowledge of their urban environment and intended audience as initiatives evolve. As such, without this ability initiatives are not able to adapt to emerging social needs or the changing properties of the urban system.

References

Adams E, Arnkil R (2013) Supporting urban youth though social innovation: stronger together Urbact II. URBACT, Brussels
Baker S, Mehmood A (2015) Social innovation and the governance of sustainable places. Local Environ Int J Justice Sustain 20(3):321–334
Fainstein SS (2010) The just city. Cornell University Press, New York
Howaldt J, Schröder A, Butzin A, Rehfeld D (2017) Towards a general theory and typology of social innovation. Technische Universität Dortmund, Dortmund
Klein J-L (2009) Territorial development and social innovation. Can J Reg Sci 32(1):3–12
MacCallum D, Moulaert F, Hillier J, Haddock SV (eds) (2009) Social innovation and territorial development. Ashgate, Farnham
Manzini E (2015) Design, when everybody designs: an introduction to design for social innovation. MIT Press, Cambridge, Massachusetts
Marcuse P (2009) From critical urban theory to the right to the city. City 13(2–3):185–197
Moulaert F, Martinelli F, Swyngedouw E, Gonzalez S (eds) (2010) Can neighbourhoods save the city? Community development and social innovation. Routledge, Abingdon
Mulgan G (2006) The process of social innovation. Innov Technol Governance Glob 1(2):145–162
Pisano U, Lange L, Berger G (2015) Social innovation in Europe. European Sustainable Development Network, Vienna

Part V
Social Innovation Model in the Production of Urban Food Spaces

Chapter 14
Social Innovation Model: Processes, Patterns, and Contextual Forces

Abstract This chapter discusses key insights in relation to the research questions, their broader meanings and implications. The underlying research question is: 'How are food growing spaces in cities co-produced between urban stakeholders at three different organisational levels of social innovation, characterised as residents, non-governmental organisations, and local authorities?' To investigate this question, a conceptual framework was formulated by drawing on studies on social innovation in urban spaces in combination with insights on innovation process (Chaps. 3 and 4). This chapter develops the framework into a dynamic model of spatial production to advance current understanding of social innovation. The sequence of innovation processes within this model is shown to have application to the development of urban food spaces both within UK cities and across Europe, while suggestions for potential research in the future are offered.

Keywords Social innovation model · Social innovation processes · Innovation patterns · Contextual forces · Spatial scales · Co-production of space · Urban food growing · Social innovation

14.1 Insights Emerging from My Research

Despite each of the cases having its own specific geographical context, situation, and characteristics, there are observable patterns emerging from my longitudinal study of social innovation process across the cases. The patterns suggest wider trends and understanding in the production of urban food spaces. I will now discuss my findings in relation to my research questions, their broader meanings, and implications.

Patterns of urban change: Observable sequence of social innovation processes in the production of urban food spaces

Based on the premise that the case studies involved collaborative action in the production of urban food spaces, there is an observable pattern to key social innovation processes. In particular, the pattern indicates an 'ideal type' sequence of social innovation in spatial production (Fig. 14.1). As all six case studies across the cities of Brighton Hove and Portsmouth, UK passed through this basic sequence (see

N. Ardill, *Growing Food in Cities*, Cities and Nature,
https://doi.org/10.1007/978-3-030-98475-5_14

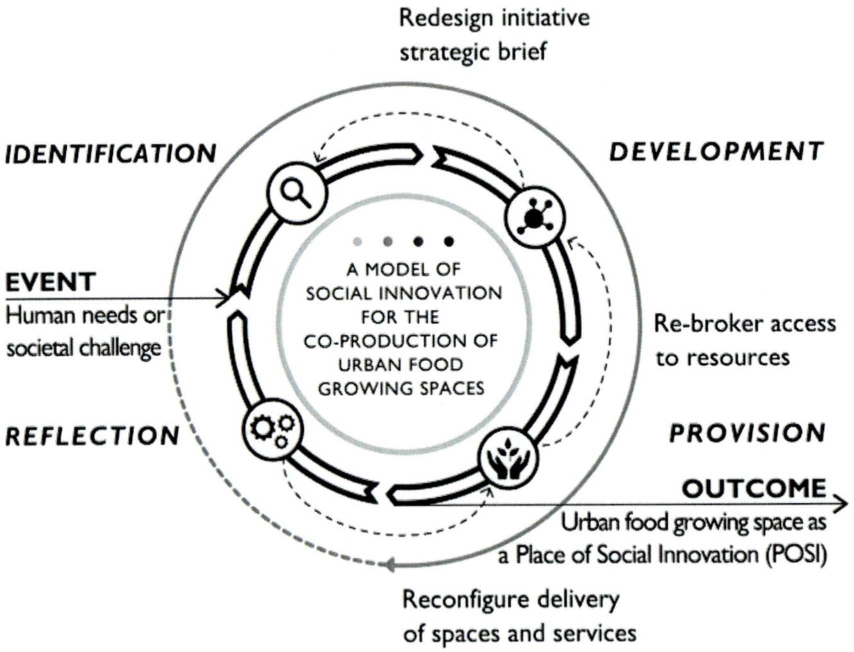

Fig. 14.1 Basic pattern in key social innovation processes for the co-production of urban food growing spaces (*Source* Author)
Note Dashed arrows between key processes indicate the revisiting of prior processes in order to progress innovation

Chap. 12), it suggests a robustness of the innovation pattern. Furthermore, the pattern can be generalised to other social innovation undertaken by these three categories of urban stakeholders (in residents, non-governmental organisations, local authorities) who are producing common spaces for food growing in cities. I do not contend that every case study corresponds to the pattern with exactness, indeed small variations from this pattern are observed. This is accounted for in challenges and opportunities experienced by focal stakeholders in working between key innovation processes. These experiences give reason for how and why the cases developed in such a manner. As such, this basic pattern draws further attention to the influence of earlier processes on those subsequently performed.

The identified pattern has significance for my study's conceptual framework as it enables the positioning of key social innovation processes in space and time. The basic pattern is supported by incorporating the 'event' (in the sum of contextual inputs) and 'outcomes' (in outputs and observable changes) as components to the key processes to support the innovation sequence.

I will now consider further the identified sequence by drawing on understandings from existing research. Eisenhardt (1989) contends that theory building from case studies and rich qualitative data can present challenges. Moreover, engaging pattern recognition across cases to generate process theory presents analytical challenges. When analysing longitudinal research on change, Pettigrew (1997) stresses 'inductive pattern recognition has also to go hand in hand with deduction' (p. 339). Equally, interpretation is seen by Langley and Tsoukas (2016) as important to give meaning to case events to connect data and theory within sequential analysis. Based on this understanding the sequence was examined against the general-to-specific readings from the cases using process theory building measures of accuracy, generality, and simplicity (Chap. 2). My observations in combination with the in-depth analysis of six case studies indicates robustness and external validity in the pattern's conceptualisation. The examination of six cases and depth of process data gained adheres to the recommendation of Van de Ven and Poole (2004) in substantiating broader conclusions about patterns and for strengthening the ability to generalise research findings. Moreover, the measurement process to determine validity and applicability, and thus the property of being in accord with fact or reality follows Hammersley (1987) in 'an account is valid or true if it represents accurately those features of the phenomena, that it is intended to describe, explain or theorise' (p. 69).

In terms of its wider relevance and practical application, the identified sequence of social innovation is suggested to be compatible with the production of urban food spaces that is happening in cities in other countries. Drawing on case studies from European cities (Chap. 5), a graphic visualisation of the process pattern is presented for four selected city examples from across different geographical regions of Europe in Figs. 14.2 and 14.3. It is important however to contextualise the identification of this innovation pattern. I have shown that the basic pattern holds a strong suitability for the three levels of urban stakeholders' action across two UK cities which were principally embarked on between 2008 to 2018. As such, this pattern might be dependent on space and time, especially on the relations among people and socio-spatial dynamics as central factors. In order to investigate the external validity of my findings it is recommended that future research should continue to examine other cities and across alternative time periods.

Another consideration of the pattern relates to the extent in which case studies represent the successes (or failures) of key social innovation processes. The varied case outcomes in terms of material spaces developed and local needs being met through services provision, or participants being socially empowered poses the question as to the necessary fulfilment of the depicted sequence to spatial production. Although it is maintained that the dependency of social innovations on space and time dynamics, specifically those taking place over longer periods in the production of space, meant that cases were liable to exhibit mixed results here in the production of urban food spaces. As such, each of the case studies suggest advancement towards the embedding of cultivation spaces within the urban fabric to varying degrees. Nonetheless, it is useful to query what might happen to their development if focal stakeholders were to instigate social action at different process stages or skip individual stages altogether? Would this cause different outcomes?

In fact the cross-case analysis reveals that different flows of social action are apparent within cycles of social innovation and phases of urban development (see Chap. 12). For example, Southsea Greenhouse, Portsmouth underwent several design

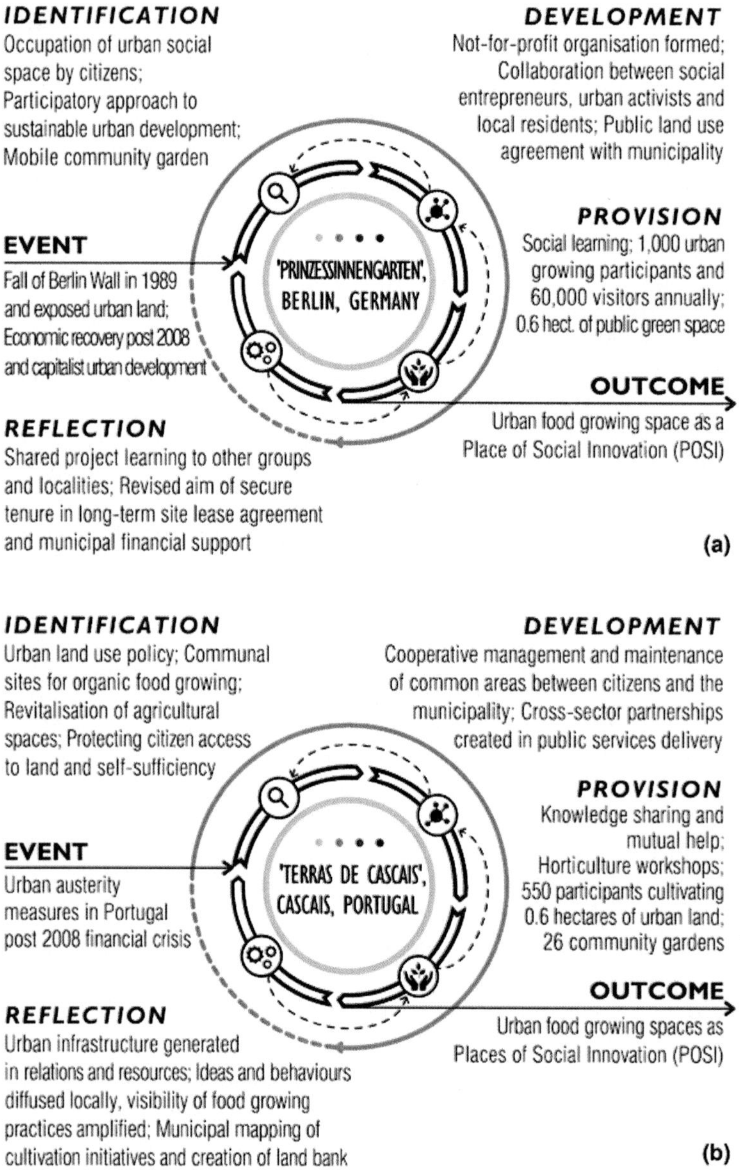

Fig. 14.2 Application of the social innovation process pattern in the production of urban food growing spaces in Western and Southern Europe cities (*Source* Author)
a *'Prinzessinnengarten'*, Berlin, Germany as an urban community food growing space that was grassroots driven. **b** *'Terras de Cascais'*, Cascais, Portugal in which several urban community food growing spaces were instigated through an urban land use policy adopted by the municipality

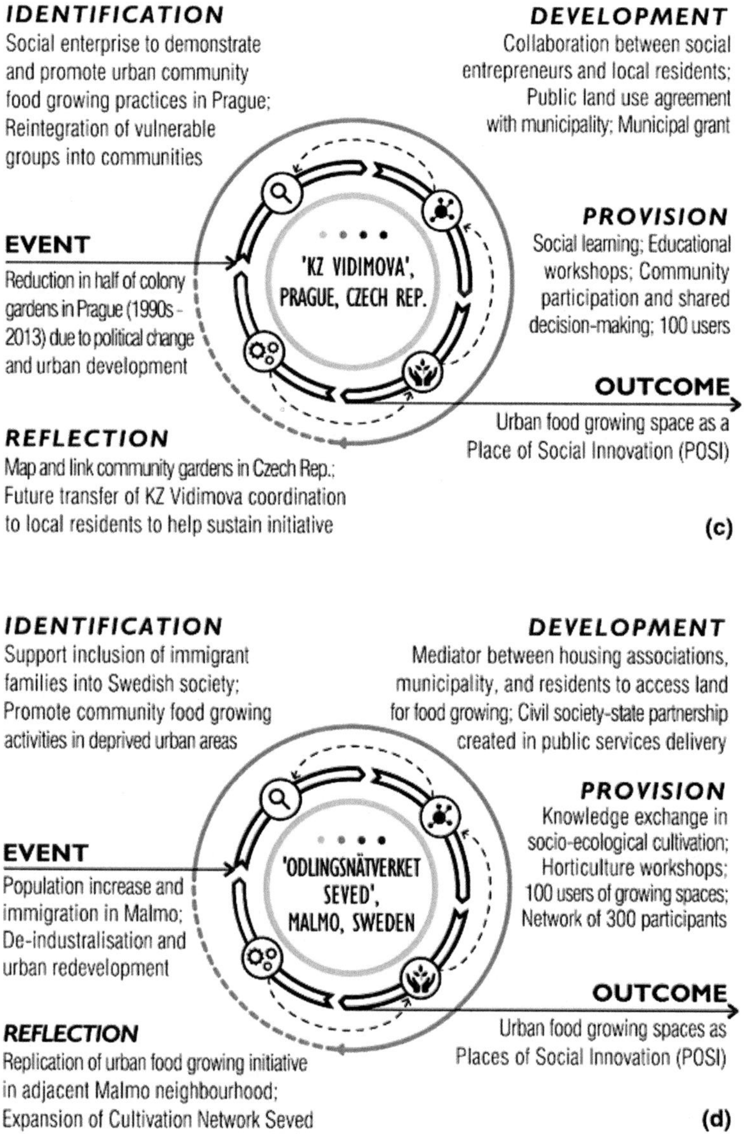

Fig. 14.3 Application of the social innovation process pattern in the production of urban food growing spaces in Eastern and Northern Europe cities (*Source* Author)
c *'KZ Vidimova'*, Prague, Czech Republic as an urban community food growing space that was instigated and coordinated by a social enterprise. **d** *'Odlingsnätverket Seved'*, Malmo, Sweden in which several community food growing spaces comprised a local network that was managed by a non-profit association on behalf of the municipality

revisions in strategies and spaces during its emergence. Would this infer that provision of opportunity to help satisfy needs can occur without prior identification of needs or the development of social relations? The initiative still needed to be designed with its intended beneficiaries in mind and resources assembled, such as people, funding, and spaces (if not externally) before provision activity could be delivered in each development cycle, e.g. offering skills workshops to local residents. It is proposed that within each cycle some identification of needs is again commenced by project stakeholders, even if limited with others, and that the co-production of space commonly dictates the development of social relations in order for these stakeholders to negotiate resources, especially material in land and funding, to implement innovation. Moreover, the cross-case analysis of key social innovation processes was undertaken based on the data gathered. In turn the pattern depicted might be applicable to other analytical scales (e.g. a household scale or a transnational initiative). My observation suggests a further avenue for research on social innovation process that could be investigated in the future.

Spatial scales and social innovation systems: Contextual forces progress and restrict the innovation actions of urban stakeholders

It is widely recognised that social innovations tend to embed prominent spatial dimensions, being routinely dependent on locality, though recognition of how, and in what ways their evolution as context dependent is underdeveloped (see Chaps. 1 and 3). To address this oversight, I examined how contextual forces affected the actions of multilevel stakeholders to produce urban food spaces. In Chap. 3 the context of social innovation was positioned at micro (neighbourhood), meso (urban), and macro (national to global) spatial scales. This developed the combination of levels of innovation and change proposed by research on Transformative Social Innovation (e.g. Avelino et al. 2014; Haxeltine et al. 2013). To position forces at work across spatial scales the case studies were examined for contextual influences prior to their categorisation. Chapter 13 established that each contextual scale affected the production of urban food spaces in a different manner. Table 14.1 outlines these multiscale forces.

The table draws attention to the sub-components of each spatial scale and the interactions between scales. It does not attempt to isolate individual forces in order to connect them to one of the four key social innovation processes as overall the forces across scales affected to some degree all the key processes. Therefore, to fully make sense of the influence of sub-components within spatial scales to the production of urban food spaces requires an understanding of how focal stakeholders interpret case events (e.g. environmental changes in opportunities or constraints). In short, contextual forces are significant though equally how focal stakeholders interpret and subsequently react to case developments also matters.

It is appropriate that I now position these underlying forces. My study examined the actions of urban stakeholders at three categorised organisational levels. The underlying forces revealed are therefore recognised as they affect the means in which multilevel stakeholders produce urban food spaces. Furthermore, these underlying

Table 14.1 Contextual forces at different scales with influence on social innovation in the production of urban food spaces

Spatial scale	Contextual force	Influence on the production of space
Neighbourhood (micro)	Interest and recognition	Community buy-in supports the local inhabitants' acceptance of spaces and initiative sustainability; restrictions and obstacles in potential disinterest or lack of participation
	Socio-political history (past events)	Local cultures and shared values influence the activation of the local inhabitants
	Experience and skill	Affects feasibility of core stakeholder to recruit and use neighbourhood participation within initiatives; influences initiative design and communication
	Existing infrastructure, resources, and assets (physical and social)	Creates resource opportunities and resolves challenges; local insight enhances neighbourhood satisfaction and spatial outcomes
Urban system (meso)	Urban politics and policy	Support and shape initiatives; provide resource opportunities to urban stakeholders, local stimulus and funding; translate national planning guidelines into urban context; build momentum behind social and spatial practices; sets planning regulation: creates local obligations, conditions, and requires negotiation; limits agency
	Professional services and collaborations	Cooperatively developing spatial experiments and innovations; overcomes urban obstacles and deficits in knowledge and skills; lack of accountability or uneven arrangements can present barriers and disempowerment

(continued)

Table 14.1 (continued)

Spatial scale	Contextual force	Influence on the production of space
	Intermediation (people and spaces)	Links people and resources; facilitates boundary crossing spaces and inclusion of marginalised voices in urban governance
	Innovative environment	Provides encouragement, indirect support and approval; builds impetus for innovation; creates social innovation platforms or ecosystem (i.e. 'innovation spaces') through which ideas are articulated, shared and developed
National system (macro)	Government policy and strategy	Strategies sets targets for societal and urban change; funding provides support to facilitate localised change; austerity politics influence local policy and citizen agency
	Funding bodies and grant-making trusts	Funding creates opportunities for community-driven development, including social enterprise and entrepreneurship
	Advocacy organisations and networks	Enable distribution of innovation knowledge; help urban stakeholders to interpret and contextualise local needs within the broader national innovation system; influences initiative aims

(*Source* Author)
Note Certain influences can both progress and restrict the development of spaces under a corresponding contextual force grouping that occurs across different phases of development, e.g. government strategies and associated funding timeframe

forces are applicable to each situation examined. There may be variations in characteristics and outcomes when examining a variety of urban instances of emerging food growing spaces within a shared period of time. Therefore, to further validate and substantiate my findings, future studies could be based in different localities and across different time periods.

The examination of specific categories of urban stakeholders also implies that additional forces influencing spatial production may have been overlooked. For example, other types of stakeholders and producers of space, including individuals, companies and organisations from the private sector who enterprise for personal profit may also have a role in the development of social innovations (as suggested by Murray et al. 2010). In this sense, accompanying or varying forces might be applicable.

Ability of social innovation to change and adapt: Adaptability is required in the development of urban food spaces to adjust to changing environments

The case studies both within the UK and across Europe demonstrated that an ability of social innovation initiatives and stakeholders to change and adapt was a necessary quality in the development of food growing spaces in cities. While initial development briefs and delivery strategies of initiatives prepared though identifying needs are important, urban environments of social action are not fixed but continuously change. Environmental interactions, especially socio-political shifts and changes in strategic direction and relations can cause not only openings but also obstacles and even stop initiatives prior to or during delivery. Consequently, adaptability within initiatives permits necessary reflection and shared knowledge building concerning urban environments, target beneficiaries, intended audiences, local institutions and networks.

Adaptability in the approaches of innovation stakeholders and their design of initiatives is a significant quality to incorporating knowledge generated via social innovation processes in order to progress development. For example, by stakeholders adjusting or redesigning components of the initiative amidst emerging conditions such as local needs, opportunities, or threats, and through cumulative understanding of social action. Testing in practice to assist learning is therefore an important quality to the sequence of 'innovation prototyping', through experimenting and evaluating of design proposals, to effect urban change. This activity enables the target beneficiaries to embed initiatives into the urban fabric and their everyday practice, i.e. for local inhabitants 'to take ownership'. However building in adaptability is a balancing act. On the one hand, initiatives should not have an overly fixed design, restricting movement and changes in values and behaviour. On the other hand, initiatives should not be inadequately designed so that fundamental aims and objectives are not clear. Recognising the characteristic of adaptability in innovation activity also has practical and research significance in expanding the production of urban food spaces. For practitioners the ability to change and adapt within and across their initiatives is a supportive recommendation, and for future research, observing innovation in the adaptability of spatial initiatives can be further investigated.

Stakeholder relationship and contact: The community relationship and interactions of stakeholders to intended beneficiaries of social innovation is variable

The case studies question and substantiate the connection between case stakeholders and neighbourhood participants; drawing the conclusion that existing research can often overemphasise its connection and constancy. The consequences of my finding become more prominent when considering the current state of knowledge on social innovation in urban spaces (Chap. 3). This existing literature often suggests there is consistency in the proximity and attachment between the social innovation initiative and the intended beneficiaries of urban change. Social innovations are typically considered to emerge and develop in a manner fitting to contextual conditions, and are characterised by participation from the bottom-up whilst oriented towards addressing the particular problems and needs of localised communities (e.g. MacCallum et al. 2009; Moulaert et al. 2010; Mulgan et al. 2007). Nonetheless, based on the evidence from the UK and European cases studied I found that community relationship and contact can sometimes be taken as a given, be unsubstantiated, and not always fully developed. Across the cases, focal stakeholders needed to continuously acquire knowledge on intended beneficiaries and contexts of social action as initiatives evolved. An absence of detailed understanding of contextual needs typically created obstacles and barriers to growth in the production of urban food spaces.

An awareness of neighbourhood interest, socio-political history, shared values, and social and spatial practice is needed to support community relationship and contact. This also involves an awareness of existing infrastructure, resources, and assets (see Table 14.1). The cases examined demonstrate that gathering a detailed understanding is not readily accomplished; it requires time and effort, and thus dedicated resources where available. Such an approach by innovation stakeholders would enable a continual reassessment and evaluation of contextual needs to ensure that implemented strategies are meeting the actual needs of local people.

14.2 Social Innovation Model for the Production of Urban Food Spaces

After considering the insights gained from my research, I now progress the conceptual framework into a dynamic model of social innovation for the production of urban food spaces. I have enabled this by drawing on 12 case studies from European cities and through an in-depth analysis of six cases within two UK urban areas of Brighton & Hove and Portsmouth over a sustained period of time (Chaps. 5–11). As such, further primary research of innovation processes, including from alternative research positions, would offer additional certainty to this framework. However, the case studies demonstrate and explain four key social innovation processes and associative components of event and outcome to support the sequence of spatial production within urban environments. The dynamic innovation model is shown in Fig. 14.4.

Social innovation action is positioned within three contextual spatial scales: neighbourhood (micro); urban system (meso); and national system (macro).

The neighbourhood scale indicates certain types and intentions of innovation activity based on the neighbourhood's inhabitants, area socio-political history, experience and skill of residents, and existing infrastructures (physical and social). Neighbourhood agency is shaped by the interrelations of urban institutions, structures

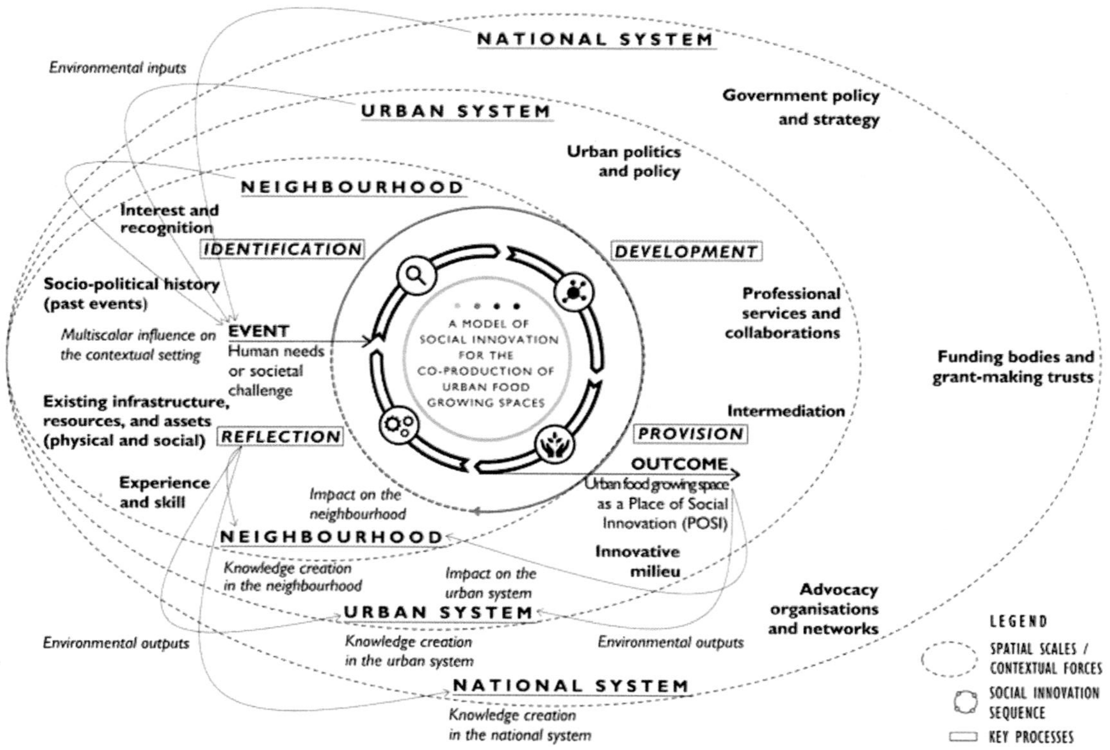

Fig. 14.4 Model of social innovation highlighting the interactions between contextual influences and key processes in the production of urban food growing spaces (*Source* Author)
Note Dashed lines in oval configuration indicate boundaries of each contextual scale: neighbourhood; urban system; national system. Long thin arrows indicate direction of environmental inputs and outputs in relation to spatial production

and systems of which neighbourhoods form a sub-component, with resident driven activity both progressed and restricted by interactions at the urban system and from the national system scale.

Socio-political and cultural changes in the urban system can prompt the identification of needs, challenges, and openings for innovation. Contextual dynamics of urban politics, professional collaborations, intermediation, and an innovative environment all influence social innovation. At the urban scale, national policies and strategies are interpreted according to specific contexts. These localised environments may create openings or barriers. Therefore, barriers and openings to innovation and the production of community food growing spaces, whether caused by the urban or national scale, are formed by urban environments.

At the national scale, government policy and strategy, funding bodies, advocacy organisations and networks function as the wider interactions of the national system to affect localised activities. Central Government strategies set goals for urban change; steering the direction of public and third sector funding whilst national networks create and share knowledge of innovation. These agendas affect the strategies of neighbourhood to urban level initiatives whilst stimulating social innovation action and establishing limitations to this action.

Regarding the structuring of geographical scales, innovation and change, Cox (1996) poses the question: 'scale seems to make a difference; but does it really?' (p. 669). By way of reply, an explanation provided by the model's conception concerns multiple geographical scales forming a shifting innovation environment for spatial production, in which different geographical scales are influencing innovation. These spatial scales have diverse interactions which positively or negatively affect the development of urban stakeholders' innovation activities. Significantly, these scales are additionally interwoven in a hierarchical structure of bounded spaces and social relations; the neighbourhood is contained by the urban system which itself is a component within the wider national innovation system. In this regard, Lefebvre (1991) reminds us that space and spatial production is subject to power relations, and thus 'each fragment of space subjected to analysis masks not just one social relationship but a host of them that analysis can potentially disclose' (p. 88). By situating processes within a broader multidimensional social and spatial framework this reveals how each spatial scale both progresses and restricts the production of urban food spaces and the actions of innovation agents.

A further explanation from this reading of scale concerns the social innovation process and the resultant environmental shifts in the neighbourhood and respective urban system. How urban stakeholders undertake social innovation in the production of spaces for food growing involves four key processes: (1) Identification, in gathering information to interpret contextual openings and constraints; (2) Development, in the external configuration of human and material resources; (3) Provision, in the delivery of public spaces and services; and (4) Reflection, in creating new knowledge. The following paragraphs present this social action in an 'ideal type' sequence of innovation.

The event is the trigger that highlights the need for social innovation. For example, this prompt may involve economic crisis, urban decline, deficient public services and facilities, endangered neighbourhood space, or lack of opportunity. The event causes social needs or spatial injustices at the neighbourhood to urban scale in combination with broader societal challenges to sustainable development. Confronted with local deficiencies in necessary social and physical infrastructure, the first stage is to recognise what action is required.

To identify needs, and significantly, emerging opportunities to satisfy these needs (via the design of innovative solutions), urban stakeholders use research to gather local insight. Focal stakeholders interpret and contextualise national strategies and agendas to their urban system. The relationship of stakeholders and contact to their target neighbourhood(s) in addition to experience and skill (internal and external) also aids generating ideas for possible solutions. Local needs should be interpreted, and then spatial interventions ideally planned in alignment within their intended context. The next stage is to approach others to gain support and develop ideas.

The development of social relations can build social networks and movements behind the proposed design of the initiative and its use within the urban system. Building relations is important to negotiating backing; human, financial, material etc. from multilevel individuals and organisations in addition to intended beneficiaries (i.e. local inhabitants) as the end users of spaces and services. In the initial phases of collaborative spatial development, the design of interventions is advanced with backers and beneficiaries whilst multilevel buy-in is sought. Intermediary agents and innovation spaces can be used to negotiate the participation of local people and external groups into plans and initiatives to enhance the legitimacy and authority of focal stakeholders, increasing access to necessary resources. Agents may act for neighbourhoods or initiatives and help develop relationships for them by connecting emerging interest for interventions, resources, and initiatives to the people and institutions of most influence.

In developing relationships, communication and negotiation can be multidirectional. Interactions between innovation participants can result in different interpretations of the spatial intervention, situation, and plans, and thus propose differing ideas regarding the necessary design of initiatives. This difference may result in modifications to the design and objectives of spatial initiatives as stakeholders must react to these varying interpretations in order to respond to or integrate alternating perspectives. In this process, co-operation between participants and neighbourhoods characteristically causes a reinterpretation of the design of interventions and local environment. The co-production of 'transformative knowledge', needed for urban transition and change, is therefore generated by local groups of innovation participants within cities. Overall, the greater an initiative's complexity (area coverage, number of beneficiaries, project stakeholders etc.) the greater the potential for project revisions required between the identification of needs and development of relations.

Provision of opportunity follows the identification of needs and development of relations, where urban stakeholders' advance initiatives to provide material spaces, and thus create opportunities for social empowerment of local inhabitants to enable satisfaction of these needs. Selected highlights from the cases of social innovation in urban food spaces in my study comprise the provision of skills workshops, trading facilities, community activities and events. A combination of delivery strategies and spaces can be employed for important elements of the initiative. For example, in order to build community capabilities the 'Brighton & Hove Harvest' urban development programme created a citywide network of cultivation projects and was used as a platform to share knowledge and experience of urban food growing (see Chap. 7). Furthermore, provision spaces characteristically involve multiple delivery strategies, rather than a singular approach, and this variety enables the underlying initiative to increase outreach and its effects on the urban environment.

Outcomes are planned or accidental, short-term or long-term, and varying in social and spatial influence. For example, project outcomes may involve generating new knowledge, creation of cultivation demonstration spaces, replication of spatial practice, and influencing decision-makers and policy to support long-term change. Moreover, innovation agency can establish new participants or social networks in the urban system, create impetus for alternative visions of urban change, though also reveal errors to growth and failures to sustain initiatives. These outcomes can be seen in changes to areas targeted for interventions and the wider urban fabric, system and structure.

Reflection of socio-spatial development practice typically closes each cycle of innovation whilst connecting existing cycles to new cycles. Learning is undertaken by urban stakeholders to generate and acquire new knowledge, whilst this process influences the identification of future needs and opportunities. Learning is also influential when unexpected difficulties occur during cycles of development. Stakeholders become aware of the limitations and constraints of the proposed solution, the initiative, and any supportive coalition developed. Attributes of the urban environment may surface or shift, affecting the realisation of provision spaces. For example, the continuation of required participation (and degree) of key stakeholder and project partners can challenge urban food growing initiatives. Evolving contextual scales and forces, such as changes in urban to national strategies and agendas during the development of the initiative, can also pose obstacles to delivery. This may require supplementary action in the identification of needs and development of relations. Moreover, the initiative might subsequently need to be redesigned and then renegotiated with key urban institutions such as the local authority and further resources obtained to realise development.

In short, the presented sequence of social innovation accounts for how multi-level stakeholders produce urban food spaces to support sustainable development. The central explanation concerns the emergence of cultivation spaces in cities as a 'spatiotemporal' occurrence, relating to both space and time. This continually shifting dynamic is influenced by the relationships of people and spaces across interconnected spatial scales and innovation systems.

References

Avelino F, Wittmayer J, Haxeltine A et al (2014) Game changers and transformative social innovation. Transit, Brussels

Cox K (1996) Editorial: the difference that scale makes. Polit Geogr 15(8):667–669
Eisenhardt M (1989) Building theories from case study research. Acad Manag Rev 14(4):532–550
Hammersley M (1987) Some notes on the terms "validity" and "reliability." Br Educ Res J 13(1):73–82
Haxeltine A, Avelino F, Wittmayer J et al (2013) Transformative social innovation. In: Social frontiers. Nesta, London
Langley A, Tsoukas H (eds) (2016) The SAGE handbook of process organization studies. SAGE Publications, London
Lefebvre H (1991) The production of space. Blackwell Publishing, Oxford
MacCallum D, Moulaert F, Hillier J, Haddock SV (eds) (2009) Social innovation and territorial development. Ashgate, Farnham
Moulaert F, Martinelli F, Swyngedouw E, Gonzalez S (eds) (2010) Can neighbourhoods save the city? community development and social innovation. Routledge, Abingdon
Mulgan G, Tucker S, Ali R, Sanders B (2007) Social innovation: what it is, why it matters, how it can be accelerated. Skoll Centre for Social Entrepreneurship, Oxford
Murray R, Caulier-Grice J, Mulgan G (2010) The open book of social innovation. Nesta, London
Pettigrew AM (1997) What is a processual analysis? Scand J Manag 13(4):337–348
Van de Ven A, Poole M (eds) (2004) Handbook of organizational change and innovation. Oxford University Press, New York

Part VI
The Progression of Social Innovation in Urban Food Spaces

Chapter 15
Social Innovation in Urban Spaces: Moving the Field Forward. A Process Framework

Abstract This chapter reflects on the research aims and findings, whilst the implications of the research are discussed. It advances the current understanding of social innovation, including the processes and contextual dynamics involved in spatial production and the contribution towards realising sustainable development.

Keywords Social innovation model · Co-production of space · Urban food growing · Policy-making · Sustainable cities · Sustainable development · Social innovation

The development of alternative approaches to innovation and sustainable cities requires further attention. As Hodson et al. (2016) stress: 'there are widespread efforts to purposively make new cities and to remake existing cities. This view is concerned not with the incremental and ongoing remaking of the city but with its transformation' (p. 128). Yet, this shared interest is not just about remaking the material fabric of the city but understanding the roles that human agency, scale and space play in its transformation. In this book I have shown that a key driver of urban change is social action, and an understanding of the process of social innovation is therefore critical to supporting strategies for realising sustainable development.

How social innovation progresses in urban environments is, however, conditional and is influenced by a variety of social and spatial dynamics, many of which have been examined within the preceding chapters. The central aim of my research was to investigate innovation processes in the collaborative production of urban food spaces in order to advance understanding of social innovation. I therefore propose that cross sector strategies are central to this approach and can be used to tackle societal challenges of increased urbanisation, health and wellbeing, social justice, food security, and climate change. At the same time these development strategies offer more localised solutions to neighbourhoods and communities. Social innovation in sustainable development therefore aims to empower all citizens through a more participative and inclusive form of urban development to meet human needs.

The book is divided into six parts. I have shown how there is a need to further understand the influence of innovation agents, associated cooperative processes, and contextual forces on urban change in the opening two parts. Additionally, I

N. Ardill, *Growing Food in Cities*, Cities and Nature,
https://doi.org/10.1007/978-3-030-98475-5_15

constructed a conceptual framework to understand this social action, using understandings from research on social innovation and insights from process research. In the third part, comprising the main body of this publication, I have tested and developed the framework by examining multiple cases of social innovation in urban food spaces. The fourth part brought together findings from across the case studies to identify a common pattern of innovation and contextual influences across spatial scales. The fifth part proposed a dynamic process model of social innovation in spatial production. This sixth and final part discusses the contributions made in the study to advancing the current understanding of social innovation. I conclude with recommendations to policymakers to support social innovation in urban food spaces.

Throughout the preceding chapters, there has been an increased awareness of the social value that civic participation and collaboration within cities offers to sustainable development strategies. This is a key argument flowing through this book and should be used by urban stakeholders interested in social innovation to support the making and organisation of sustainable cities.

15.1 Advancing Current Understanding of Social Innovation in Spatial Production

I have presented in this book a unique understanding on the process of social innovation in the production of urban food spaces. The innovation model proposed is advantageous not only to urban planners and policymakers but also to practitioners as it delineates how stakeholders develop food growing initiatives and the interactive and multiscale innovation environment they must operate within. Furthermore, it responds to Cajaiba-Santana's (2014) call for a process framework that enables deeper understanding of the complexities of social innovation agency undertaken in time-based and environmental dependent contexts, as well as the call of Moulaert and Mehmood (2011) for more sector-specific research to integrate findings on the spatial characteristics of social innovation in order to advance conceptual understanding (Chap. 1). My proposed model is a localised model in contrast to the global model of transformative social change offered by Haxeltine et al. (2013), especially in the causal interactions between context specific social innovations and systemic and structural social change.

The innovation model advances understanding on the evolution of social innovation to explain how productive green spaces are collaboratively developed by different groups of people in different urban contexts. Through determining the various factors that affect the development of social innovation initiatives it contributes a multi-dimensional view of urban change. This dynamic model integrates geographical aspects (environment, infrastructure, resources) and sociocultural aspects (interest, experience, support) in addition to politics and policy. Furthermore, it advances the understanding of social action in urban change by integrating innovation systems and interactive processes at hierarchical spatial scales.

Through the formulation and testing of this model, I have established an 'ideal type' sequence to social innovation processes. The production of urban spaces passes through this sequence; an event triggers human needs, subsequently these needs are identified by urban stakeholders who develop relationships with others to access resources, next provision activities emerge to satisfy needs, causing outcomes in the production of space, lastly reflection and learning occurs in the creation of development knowledge. I propose that this sequence can be generalised and applied across urban stakeholders' production of cultivation spaces to cities both within the UK and across Europe (see Chaps. 12 and 14). As such, the sequence advances the framework on key social innovation processes developed from current social innovation research. This has been achieved through an in-depth analysis of three organisational levels of urban stakeholder action across a sustained period of time, thereby setting this activity both contextually and chronologically.

In addition, I have categorised the various contextual influences on the production of space. The identification of three spatial scales of social innovation is an important result of my study and develops a multilevel understanding of change from innovation studies and my in-depth analysis of case studies. A reasoned mapping of contextual forces through this approach is also important for the expanding interest in social innovation for sustainable development as it provides a more layered understanding of context in social innovation process. In highlighting the significance of this contribution, Moulaert et al. (2017) has stressed that the current lack of spatial understanding has presented challenges to innovation growth. As such, my observation tests and strengthens the view that social innovation systems comprise of multiple scales and are nested within spatial hierarchies. Overall, my research findings substantiate the lack of detail in current explanations, expressly the 'process' through which social innovations gain transformative impact, to explain the agency required to provide opportunities to satisfy human needs and tackle societal challenges.

15.2 Reflections on Current Debates Surrounding the Meaning of Social Innovation

Recognising key social innovation processes in urban change is an important contribution to debates within social innovation research, especially on-going discussions about conceptual meaning. I have shown that social innovation does not develop from a 'power vacuum'; the process is interwoven and interdependent in governance processes across multiple scales, structures and conditions that frame a shifting innovation environment. As argued within the review of existing research (Chap. 3), the emergence of social innovation is often attributed to grassroots movements external to government, typically in response to state shortcomings in meeting social needs (e.g. Moulaert et al. 2010, 2013). I have extended the understanding on social innovation by concluding that urban transformations around the development of common spaces for food growing are the product of groups of 'change agents' across public, private, and third sectors. These agents will often cooperate and collaborate in grant programmes and government-funded initiatives that create a window of opportunity for social innovation.

As such, my finding shows that social innovation does not only emerge to tackle a specific socio-spatial problem or threat, i.e. social innovation as 'problem-solving' as suggested by Moulaert et al. (2010), but also because of the opening for social action through access to necessary public resources from institutional levels. I therefore agree with the process perspective offered by Cajaiba-Santana (2014) in that social innovation is a collective and multilevel phenomenon with an interactive relationship between social agent and structure across micro, meso, and macro levels of analysis. As shown here, social innovation emerges and develops from interactions between the various stakeholders within each case and their socio-institutional context, with social institutions (e.g. local authorities) both advancing and restricting the emergence of innovation. Moreover, these case stakeholders undertook social innovation in their environment while contributing to the evolution of social systems and institutions. My study therefore confirms and extends the socio-spatial relationship understood by Lefebvre (1991) and Soja (2010) in that social action in spatial production is being shaped by the environment and at the same time itself is shaping the environment and the intensity of its social relations and interactions.

In terms of spatial relations and power, the findings of my study have implications to key debates and questions over 'radical' versus 'complementary' interpretations of social innovation (see Chap. 3). Specifically, whether the process needs to be a driver of structural changes in power relations and thus increase citizen empowerment (e.g. Moulaert and MacCallum 2019), or whether it is compatible to exiting power relationships, recognition and distribution frameworks (e.g. Mulgan 2019). The former, radical perspective puts into effect the grand revolutionary arguments of Lefebvre (1991) and Harvey (2009) on socially just urban transformation by claiming social innovation challenges existing structures, affects new institutions, and causes disruptive changes at the systemic level, as argued for by Westley and Antadze (2010). Yet it is concluded here that the embedding and institutionalising of social innovation does not have to radically disrupt existing governance structures, rather innovation needs to be integrated within these structures with spaces and conditions created for alternative visions of sustainable development to endure and reproduce through replication and diffusion.

My research complements the critical urban thinking around spatial restructuring presented by Brenner (2019) and in particular, it questions the long-standing view of Harvey (2009) on socially just urban transformations not being possible without radical system change to affect existing structures due to the institutional dependence upon past ways of working. In the cities examined, both within the UK and across Europe, change was realised within existing systems and governance structures with pressure often applied both within government and from civil society in directing urban policy, suggesting the importance of context in local agency and institutional openness to change. Here, the approach to developing capabilities (and thus the capacity to act) in the production of common spaces for food growing corresponds to the 'just city' planning advocated for by Fainstein (2010) in mobilising a broad base of participants as agents of change to press for more progressive policies. To use a term coined by Gorz (2013), these cities activated the dynamics of urban change through 'non-reformist reforms', where alternative institutional designs emerged to influence the self-determination of the urban population over the means of spatial production. Moreover, in supporting the complementary view on social innovation (e.g. Mulgan 2019), local communities here effectively collaborated with urban institutions in developing cooperative relationships within existing governance frameworks to cumulatively affect change and thus create the conditions on which to pursue longer-term urban transformations.

Lastly, the observed relationships between social innovation and the localised embedding of rules and new behaviour patterns indicates the influence of micro-relationships and especially the role of micropolitics within 'urban everyday politics' as proposed by Beveridge and Koch (2019). Micropolitics as a driver of urban change, can be explained here as small-scale interventions at the neighbourhood level that support the governing of behaviour of urban inhabitants. These urban food growing initiatives helped to shape the preferences, attitudes, and perceptions of local inhabitants by introducing the necessary skills and capacities to adapt their practices to further their development knowledge and meet their needs. This meant that traditional functions of decision-making and resource distribution were being devolved from the local authority to other levels of urban organisation to affect how individuals and social groups adjusted their behaviours within urban environments towards spatial production. Seen in this light, the creative process of informal and formal empowerment to achieve social goals suggests scope for experimentation and innovation in an environment of constant transformation. As social innovation increasingly gains prominence within political spheres, I suggest it is therefore valuable to consider the relationship of micropolitics to innovation emergence in order to further understand its influence on spatial change at the urban to national scale.

References

Beveridge R, Koch P (2019) Urban everyday politics: politicising practices and the transformation of the here and now. Environ Plan D Soc Sp 37(1):142–157

Brenner N (2019) New urban spaces: urban theory and the scale question. Oxford University Press, New York

Cajaiba-Santana G (2014) Social innovation: moving the field forward. A conceptual framework. Technol Forecast Soc Change 82(1):42–51

Fainstein SS (2010) The just city. Cornell University Press, New York

Gorz A (2013) Capitalism, socialism, ecology. Verso, London

Harvey D (2009) Social justice and the city, revised edn. University of Georgia Press, Athens

Haxeltine A, Avelino F, Wittmayer J et al (2013) Transformative social innovation. In: Social frontiers. Nesta, London

Hodson M, Burrai E, Barlow C (2016) Remaking the material fabric of the city: "alternative" low carbon spaces of transformation or continuity? Environ Innov Soc Transit 18(1):128–146

Lefebvre H (1991) The production of space. Blackwell Publishing, Oxford

Moulaert F, MacCallum D (2019) Advanced introduction to social innovation. Edward Elgar Publishing, Cheltenham

Moulaert F, Mehmood A (2011) Spaces of social innovation. In: Pike A, Rodriguez-Pose A, Tomaney J (eds) Handbook of local and regional development. Routledge, Abingdon, pp 212–225

Moulaert F, Martinelli F, Swyngedouw E, Gonzalez S (eds) (2010) Can neighbourhoods save the city? Community development and social innovation. Routledge, Abingdon

Moulaert F, MacCallum D, Mehmood A, Hamdouch A (eds) (2013) The international handbook on social innovation. Edward Elgar, Cheltenham

Moulaert F, Mehmood A, Manganelli A (2017) Social innovation spaces. In: Monteduro G (ed) Welfare, subsidiarity and social innovation. FrancoAngeli, Milan, pp 71–93

Mulgan G (2019) Social innovation: how societies find the power to change. Policy Press, Bristol

Soja E (2010) Seeking spatial justice. University of Minnesota Press, Minneapolis

Westley F, Antadze N (2010) Making a difference: strategies for scaling social innovation for greater impact. Innov J Public Sect Innov J 15(2):1–19

Chapter 16
Supporting Social Innovation in Urban Food Spaces and Strategies for Sustainable Development

Abstract This concluding chapter makes recommendations to policymakers, planners, and others interested in urban food spaces and sustainable development to further social innovation in urban change. The chapter ends with some final remarks on the socio-ecological value presented by social action in the production of urban food spaces.

Keyword Social innovation model · Co-production of space · Urban food growing · Policy-making · Sustainable cities · Sustainable development · Social innovation

16.1 Recommendations for Policymakers

The findings from my research have additional relevance to policymakers, planners, and others interested in sustainable development strategies. In formulating the social innovation model, my aim was to gain understanding on the production of urban food spaces. The proposed model builds upon a lack of current knowledge and policy-oriented awareness of urban change to explain the process of social innovation as 'spatiotemporal' dependent, in relating to both space and time.

The sequence of social innovation presented in this book offers a general explanation of the actions involved in producing urban food spaces. For policymakers an increased awareness of this collaborative spatial production emphasises the significance of various 'change agents' across civil society, public, and private sectors as performing an important function in shaping innovation and the emergence of urban food spaces. This emphasises that social innovation is connected to, as opposed to being separate from, the activities of additional urban participants.

Understanding the production of common spaces for cultivation within urban environments as being performed within three contextual spatial scales has an additional importance for urban to national decision-makers. The model highlights how government strategies, policies, and agendas are significant functions in shaping national innovation systems, and yet this broader environment is continuously moderated by urban environments. Therefore, while national development strategies and policies are significant, this means that urban politics and policies, organisational support,

N. Ardill, *Growing Food in Cities*, Cities and Nature,
https://doi.org/10.1007/978-3-030-98475-5_16

and an innovative local environment are also significant functions. As such, this awareness opens up a space for innovation in which alternative pathways to urban transformations can arise, in which strategies of place-based change can perform a leading function.

The outcomes of my research have led me to make the following recommendations to urban policymakers in order to support social innovation strategies in sustainable development:

Create a professional role for a social innovation coordinator at the city administration level to support the production of urban food spaces:

Social innovation coordination would be a valuable strategy at all spatial scales, especially at the urban and neighbourhood scale. The process of social innovation in spatial production is dynamic, suggesting that the skill and function of a coordinator will require flexibility, creativity, and the ability to influence development. For example, coordination would mobilise and educate cross-sector stakeholders to build a sense of collective agency and shared responsibility to their built environment. This process would facilitate local communities to have influence over issues that impact their quality of life whilst helping government institutions to work more effectively in meeting social needs to improve wellbeing. To address local needs, the new professional could help to identify and confirm the areas of service provision requiring intervention, such as public access to cultivation spaces. Moreover, such a coordinator could help innovators at all levels to secure the necessary funding and resources for initiatives whilst removing barriers and obstacles, such as resource deficits in knowledge and skills, in order to resolve challenges during planning and negotiation stages of initiative development.

What is evident from all of the case studies examined in this book is that there is potential for increased agency of local authorities in coordinating food growing projects on public land. The appointment of a responsible officer would provide city-wide and institutional support to local networks of food growing spaces and their practitioners, whilst creating additional opportunities for green space development. This officer could serve as an intermediary and primary point of contact for local people and organisations in coordinating external approaches to other relevant officers within the local authority. For instance, in Portsmouth, UK a local authority officer was a key intermediary between an environmental organisation and the City Council in facilitating access to institutional support and the co-design of several public green spaces for food growing.

Local authority officers with a personal interest in food growing and civic collaboration are ideally positioned to help build trust and relationships between the City Council and local residents. An ongoing state-civil society interaction will support an identification and understanding of the contextual opportunities and barriers to developing localised food spaces. This knowledge can be fed back into the local authority and permit flexibility and solution-oriented interpretations of planning policy and legislation, for example in relation to the activation of vacant public spaces or temporary use of construction sites for community food growing. Oversight at the strategic level would also ensure access to relevant information from various local authority departments and an increased influence over public policy and decision-making.

In terms of finding the appropriate model for organisation within the local authority, this may involve an individual coordinator, or a task force that operates across the City Council and who have close contact with civil society groups. An alternative approach would be for the City Council to partner with a local organisation with existing community links and work with them to allocate urban resources and coordinate community food growing. This partnership model is successfully used in Brighton & Hove, UK where the non-governmental organisations Brighton & Hove Food Partnership and Brighton Permaculture Trust have co-managed the local development of over 50 new community food growing spaces.

Provide opportunities to local people and community groups in the ongoing management and maintenance of urban food spaces:

Many participants of citizen-led food growing initiatives are motivated in the creation phase of an urban community garden when it is novel and energising. When challenges of retaining participants occur, whether through lack of interest or internal group conflicts, the availability of organisational support from the local authority (or an external partner organisation) can help to continue activities by maintaining the commitment of those involved. An understanding of food growing initiatives and their participants can be used to allocate stakeholder roles and responsibilities with an adaptable set of guidelines and procedures to ensure that participants do not disappear due to a lack of governance.

Community food growing spaces should ideally create opportunities among participants for social empowerment, a key objective of social innovation initiatives. Civic participants therefore benefit from a culture of collective ownership and decision-making over neighbourhood spaces and activities, especially as their active engagement in green space management can enhance community capabilities and improve project sustainability. This creation of social value is evident in Utrecht, the Netherlands where the local municipality has developed a novel strategy for its public spaces that is based on the collaboration with citizens. The strategy promotes civic participation in the greening of the city and its development to facilitate opportunities that will improve the quality of life for its citizens. This process has been stimulated through the liberal approach of the municipality of Utrecht to participatory development and the co-management of public spaces by local inhabitants.

The enabling of cooperative processes requires shared learning to develop knowledge amongst project stakeholders and the investment of public resources in the capacity building of civil society. This approach will help to facilitate equitable participation. Inclusive forms of urban governance over neighbourhood green spaces would help to activate and involve more local inhabitants and groups (especially those not already active or adequately represented) in formal decision-making procedures, such as the creation of neighbourhood citizens' assemblies and participatory budgeting to co-design food growing spaces and services. Moreover, the right of citizens to participate in public affairs is fundamental to democratic development. This is emphasised in the 2030 Agenda for Sustainable Development, which calls for participatory and representative decision-making at all levels to achieve progress on socially just and inclusive societies (United Nations 2015).

Local government support for urban food spaces needs to be realised as an ongoing investment in people and spaces. A major challenge to the sustainability of community food growing initiatives is that funding is largely designated to the start-up phase. Ensuring viability in the longer-term, may require further public subsidies in land, materials, water supply, and importantly, remuneration of any coordinators attached to the operation of the initiative. In Portsmouth, UK for example, a significant amount of Central Government funding from the UK 'Healthy Towns' programme (2008–2011) was spent by the local authority on the creation of several community gardens in the city. This concentration on the start-up phase of development meant an uneven distribution of funds, including the retention of garden coordinators not being budgeted for, threatening further garden development.

Create a supportive network that brings people and projects together working on social innovation in urban food growing:

Additional and more sustainable urban food growing spaces can be innovated through cooperation and partnerships between local institutions, organisations, social enterprises, and networks of expert practitioners and citizens. Community initiatives should ideally be based on a detailed understanding of the local needs and conditions of area interventions including: type of users, social background, environment, etc. The social value gained will be enhanced through developing project strategies aimed at target users alongside publicly accessible spaces where additional users can also interact and share in food growing practices. The incorporation of external advice through establishing links with experts in horticulture, food, health, and education will further add value to social innovation initiatives, especially to citizen-led projects. Drawing on expert and local knowledge through the creation of stakeholder networks will contribute experience and insight. At the same time this will increase recognition, both political and cultural, whilst creating support for urban food growing by recruiting members as a collective within a local network.

The promotion of spatial interventions in the areas of social innovation, urban food growing, and sustainable development will help to create links between local stakeholders and their social agendas. This can also reduce potential competition over urban resources, such as land, funding, civic participants, and top-down support. These partnerships are ideally positioned to integrate social and public policies with service provision and civil society projects, whilst connecting the urban context to regional, national, and international strategies. For example, the Brighton & Hove Food Partnership as part of the UK Sustainable Food Cities Network represents an archetype for coordination and intermediary agency, since it brings together a range of urban stakeholders (e.g. community groups, environmental organisations, educational institutions, public health bodies, and local authorities) in order to socially innovate specific actions or support the initiatives of its members. Their combined agency within a larger network is the primary added value of this partnership, and fundamental to their way of working.

Establishing a local network between the various cultivation initiatives in a city will help to increase awareness and citizen participation whilst furthering collaboration and communication between initiatives. This can involve a mapping of existing projects on the local municipality website, such as in Cascais, Portugal where food growing sites around the city are plotted as a platform between residents looking for land to cultivate and landowners. An alternative is a non-governmental organisation working on behalf of the City Council as in Prague, Czech Republic (e.g. *Kokoza*) and Brighton & Hove, UK (e.g. Brighton & Hove Food Partnership) where community food growing initiatives are promoted and connected within wider networks. The land mapping can show all local food growing spaces with descriptions and contact information for the initiatives. Additionally, in order to utilise vacant public land and cultivate different types of urban spaces, the mapping of existing and potential spaces within a city for food growing is proposed. In Portsmouth, UK for example, the city's planning department has assisted the development of productive green spaces by mapping possible areas in the city for food growing, making it easier for community groups to obtain access and permission to use public land. This mapping activity also supports the design of initiatives by identifying areas of contaminated land and other constraints to development.

Recognise the varying power dynamics associated with civic participation, co-production, and partnership strategies to social innovation:

There has been growing attention towards social innovation and the potential of collaborative partnerships between citizens, non-governmental organisations, and statutory authorities (see, e.g. Bovaird and Loeffler 2012; Boyle and Harris 2009; Voorberg and Bekkers 2018). These publications highlight the undeveloped utility of collaborative approaches for city shaping and public service provision, especially for enhancing the effectiveness and efficiency of urban outcomes by responding to citizens' needs. However they also stress the challenges involved. It is found here that, in practice, these collaborative relationships in the production of urban food spaces will vary in their 'equality and reciprocity' (qualities seen as key to co-production approaches). It should not be automatically assumed that citizens and project partners are able to assume, or indeed desire, shared responsibility in the planning, design and delivery of urban spaces and services.

Urban stakeholders and their project partners are likely to possess varying levels of knowledge and (professional) expertise, whether between civil society, public or third sector, and this relationship can also be unequal regarding access to resources, authority, and decision-making capability. While co-production approaches suggest an opportunity to multiple resources and meet knowledge deficits, the findings of my study concur with Brandsen et al. (2016) and Moulaert et al. (2017) in cautioning against social innovation approaches to citizen participation and state-civil society partnerships being exploited for short-term gains, e.g. political or symbolic, or act as an outward sign of achievement. For instance, in Barcelona, Spain the municipal '*Pla BUITS*' plan encouraged the participation of local inhabitants in the revitalisation of the urban fabric by making vacant spaces available for public activities, such as community food growing. However, these open spaces were restricted to a maximum use of three years and, as a result, the approach was directed towards temporary urbanism rather than on long-term planning. In short, knowledge created through collaborative working should be incorporated into future planning and development proposals, and ideally, suitable degrees of stakeholder involvement and levels of responsibility tailored to specific built environments.

Underwrite social innovation as an intrinsic yet specified component of urban planning and development policy:

Social innovation is positioned as possessing a marked potential for sustainable development and consequently is seen a significant driver to transforming in the way in which people and societies produce and consume urban space. In addition to an ability to operate in-situ, connect the everyday with policy and social movements, and activate local communities, the approach typically involves urban stakeholders with shared values, beliefs, and attitudes. This allows for greater social action than under current approaches to the production of the built environment. As such, social innovation symbolises a marked capability to contest common practices, transform existing relations between social and urban change, and creates opportunities for experimentation with alternative approaches to sustainable development.

Social innovation in the production of urban spaces connects with and requires the backing of various urban to national participants across civil society, public and private sectors. Innovation can be progressed through national strategies and political agendas though also restricted by political volatility. Therefore, acknowledgement of the role of social innovation to achieving sustainable development, such as within the European Union's Europe 2020 Strategy is an important move. In order to continue to promote socially innovative agency in sustainable cities, policymakers should position this innovation as a specific component of any prospective development policy, rather than an external area of social action that might support realisation of sustainable development goals.

Consider the influence of regulation on social innovation and urban green space interventions and whether there is a need for greater planning regulation:

Regulation can stimulate urban space interventions, such as in the formation of sustainable city frameworks seen in Reykjavik, Iceland and Brighton & Hove, UK. These frameworks integrate food growing within planning and development to increase local production and visibility of productive green spaces among citizens. Alternatively, this recent growth can be observed in the collaborative approaches to public services, particularly in the development of urban green spaces through state-civil society partnerships to improve the well-being of citizens as seen in Malmo, Sweden, and in Prague, Czech Republic, and Portsmouth, UK. There is a need for regulation that facilitates experimentation and innovation in meeting the needs of innovators whilst still safeguarding the broader needs of citizens and public institutions. Regulation is however challenging and suggests a balancing act in creating opportunities for innovation to grow and accelerate whilst at the same time protecting mainstream interests. In order to balance potential conflicts and enable flexibility in regulatory approaches to innovation, the public sector can support innovation by equipping stakeholders with the suitable policies and instruments for efficient planning and delivery.

Policy approaches to the development of community food growing spaces as a key component of urban green infrastructure emphasises the relationship between planning regulation and innovation. In the UK, the Small Holdings and Allotments Act 1908 imposed a duty on local authorities to provide and let allotments to urban inhabitants for food production and these spaces are still afforded robust protection under The Allotments Act 1925, as they cannot be disposed of without Ministerial consent. However, community gardens and orchards are not equally integrated within national land use policy and planning frameworks, challenging their long-term continuation. Recognition within city development plans is variable with the production and protection of these community spaces being dependent on local rules and circumstances. As publicly accessible green spaces, I recommend that community gardens and orchards should require greater legal protection and appropriate designation within local to national planning frameworks. This means that the interests of their users and beneficiaries would need to be considered at any public review for planned changes to land use. Institutionalised support within open space planning and policy will help to safeguard the continuation of community food growing spaces as significant community assets and their important function within the urban fabric.

The use of standard agreements for the temporary use of urban land for food growing as an interim measure may present a pragmatic solution to planning regulation. Community food growing initiatives, and especially those in centralised urban locations, would gain from a standard right of use agreement over vacant public land. This agreement can outline guidelines in relation to material development, for example planting in the ground or in containers, land maintenance, public access, and notice of termination. Standardised contracts would provide initiatives with security of tenure in the short term, based on contracts of 3–5 years. If the initiative becomes established, an argument could be made to (or by) the local authority to extend this agreement or to be offered another suitable public location within the neighbourhood. This could again be on a temporary basis or transition to becoming more 'permanent' (i.e. over 10 years) depending on local circumstances. In Brighton & Hove, UK for instance the use of standard agreements between the City Council and community groups has become widespread and is now a valued model in the co-management of urban food growing spaces.

Assist the realisation of alternative development pathways to further the collaborative production of urban food growing spaces:

As a final recommendation, with public administrations facing monetary constraints to achieving the goal of sustainable development in combination with reduced budgets for welfare provision to satisfy social needs, it is necessary to acknowledge and back the formation of alternative development pathways. Urban politics and policies can steer citizen interest in collaborative city shaping, including the development of food growing spaces, whilst increasing public access to assets and common resources that are essential to the growth of social innovation. Where urban areas are seeking to promote social innovation, this should be supported through the restructuring and prioritisation of funding and social investment to encourage socially orientated enterprise and entrepreneurship.

16.2 Final Remarks: The Socio-ecological Value of Social Innovation in Urban Food Spaces

In this book I have shown how the support of social innovation in the production of urban food spaces presents authorities with the opportunity to create socio-ecological value in many ways. The preceding chapters highlighted the ways in which these spaces can function as meeting places to strengthen communities whilst enabling residents to influence their local environment by actively participating in urban development. Community food growing initiatives can be used as a framework for wider social projects and socio-economic enterprises, in addition to contributing to public health and wellbeing through promoting physical activity, contact with nature, and raising awareness of local food production and ecological processes in urban environments. As public green spaces they can support strategic objectives around urban and neighbourhood renewal in creating social, cultural, and physical development in urban areas. Furthermore, these spaces can form a significant component of urban green infrastructure and sustainable development strategies due to their environmental contribution to objectives such as increased biodiversity, carbon sequestration, urban cooling, diversion of rainwater, and improved air quality. By making social innovation and urban food growing a politically prioritised area in local strategies and development plans, the resultant socio-ecological value can be fully realised for the benefit of local people, communities, and authorities.

References

Bovaird T, Loeffler E (2012) From engagement to co-production: the contribution of users and communities to outcomes and public value. Voluntas 23(4):1119–1138

Boyle D, Harris M (2009) The challenge of co-production. NESTA, London

Brandsen T, Cattacin S, Evers A, Zimmer A (eds) (2016) Social innovations in the urban context. Springer, London

Moulaert F, Mehmood A, MacCallum D, Leubolt B (eds) (2017) Social innovation as a trigger for transformations. European Commission, Brussels

United Nations (2015) Transforming our world: the 2030 agenda for sustainable development. United Nations Environment Programme, New York

Voorberg W, Bekkers V (2018) Is social innovation a game changer of relationships between citizens and governments? In: Ongaro E, Van Thiel S (eds) The Palgrave handbook of public administration and management in Europe. Palgrave Macmillan, London, pp 707–725